Principles of AI Governance and Model Risk Management

Master the Techniques for Ethical and Transparent AI Systems

James Sayles

Apress®

Principles of AI Governance and Model Risk Management: Master the Techniques for Ethical and Transparent AI Systems

James Sayles
Spring, TX, USA

ISBN-13 (pbk): 979-8-8688-0982-8   ISBN-13 (electronic): 979-8-8688-0983-5
https://doi.org/10.1007/979-8-8688-0983-5

Copyright © 2024 by James Sayles

This work is subject to copyright. All rights are reserved by the Publisher, whether the whole or part of the material is concerned, specifically the rights of translation, reprinting, reuse of illustrations, recitation, broadcasting, reproduction on microfilms or in any other physical way, and transmission or information storage and retrieval, electronic adaptation, computer software, or by similar or dissimilar methodology now known or hereafter developed.

Trademarked names, logos, and images may appear in this book. Rather than use a trademark symbol with every occurrence of a trademarked name, logo, or image we use the names, logos, and images only in an editorial fashion and to the benefit of the trademark owner, with no intention of infringement of the trademark.

The use in this publication of trade names, trademarks, service marks, and similar terms, even if they are not identified as such, is not to be taken as an expression of opinion as to whether or not they are subject to proprietary rights.

While the advice and information in this book are believed to be true and accurate at the date of publication, neither the authors nor the editors nor the publisher can accept any legal responsibility for any errors or omissions that may be made. The publisher makes no warranty, express or implied, with respect to the material contained herein.

>Managing Director, Apress Media LLC: Welmoed Spahr
>Acquisitions Editor: Shivangi Ramachandran
>Development Editor: James Markham
>Project Manager: Jessica Vakili

Distributed to the book trade worldwide by Springer Science+Business Media New York, 1 New York Plaza, New York, NY 10004. Phone 1-800-SPRINGER, fax (201) 348-4505, e-mail orders-ny@springer-sbm.com, or visit www.springeronline.com. Apress Media, LLC is a California LLC and the sole member (owner) is Springer Science + Business Media Finance Inc (SSBM Finance Inc). SSBM Finance Inc is a **Delaware** corporation.

For information on translations, please e-mail booktranslations@springernature.com; for reprint, paperback, or audio rights, please e-mail bookpermissions@springernature.com.

Apress titles may be purchased in bulk for academic, corporate, or promotional use. eBook versions and licenses are also available for most titles. For more information, reference our Print and eBook Bulk Sales web page at http://www.apress.com/bulk-sales.

If disposing of this product, please recycle the paper

My dearest Lisa,

Words cannot express my gratitude for your unwavering support throughout this book-writing journey. You believed in me from the very first word, even when I doubted myself. Your patience, encouragement, and those countless late-night cups of coffee were the fuel that kept me going.

This book would not exist without your love and support. You were my sounding board, first editor, and biggest cheerleader. Thank you for sharing this incredible adventure with me. I love you more than words can say.

With all my love,

James

Table of Contents

About the Author xv

Preface xvii

Acknowledgments xxxi

Chapter 1: The Current State of AI Governance and Model Risk Management 1

    Current Maturity Levels on AI Governance and Model Risk Management 8

    Assessment of the Proficiency Levels of AI Governance Strategy Experts 10

    What Are the Big Four and Other Strategic Consulting Firms Doing to Address the Gap in Skills? 11

    Summary and Thoughts 14

    Quiz 15

Chapter 2: Aligning AI Strategy and AI Governance 19

    An AI Strategy Without AI Governance Is an AI Failure 19

    Refining Your AI Strategy for Principled Performance 28

    Human-AI Collaboration 28

    Generative AI and GPTs 30

    Key Strategies for Using Generative AI and GPTs 37

    Should I Use Generative AI or GPTs for Critical Business Operations? 40

    Implementing the AI Strategy 42

    Identifying and Managing the Risks of AI Strategy Implementation Failures 44

    Defining Responsibility for AI Systems 46

    Inclusive Design of AI Systems 48

TABLE OF CONTENTS

    Cultural and Social Impacts of AI 50

    Public Awareness and Education 52

    The "Best" AI Strategy Frameworks and Tools in the Business 54

    How Should AI Strategy and AI Governance Coexist? 57

    AI Strategy Cost Factors and Drivers 60

    Summary and Thoughts 63

    Quiz 64

Chapter 3: How to Sound like an AI Governance and Model Risk Management Guru 67

    What Is AI Governance? 67

    What Is Model Risk Management and Model Risk Governance? 72

    AI Governance and Model Risk Management Drivers 74

    Where Does Data Science Fit Within AI Governance? 76

    Summary and Thoughts 81

    Quiz 81

Chapter 4: Designing a Well-Governed AI Lifecycle Model 85

    AI Development Lifecycle Stages 86

    AI Model Proposal and Feasibility 87

    Identify Inherent AI Risks and Define Preliminary Metrics 89

    Model Development and Testing 91

    Model Review and Validation 96

    AI Model Approval 100

    AI Model Deployment and Implementation 103

    Ongoing Monitoring of AI Models 106

    Summary and Thoughts 108

    Quiz 109

TABLE OF CONTENTS

Chapter 5: Aligning AI Governance with Other Internal Governance Models for Trustworthy AI: "The Convergence of Governance Frameworks" 113

Establish AI Governance 114

AI Ethics and Governance Strategy 115

Data Governance 120

IT and Cybersecurity Governance 134

Managing Cybersecurity Risks AI Systems Face 147

AI Model Risk Management 151

AI Model Risk Process Recommendations 163

AI Governance Metric Definition and Reporting 164

AI Governance Key Performance Indicators (KPIs) 167

Summary and Thoughts 170

Quiz 170

Chapter 6: Designing Your AI Governance Framework 173

Frame the Goals of Building a Successful AI Governance Framework 173

A Practical Guide for Building Your AI Governance Framework 175

AI Governance Framework Model and Design Concept 179

Summary and Thoughts 180

Quiz 181

Chapter 7: AI Governance and Oversight Model 183

AI Governance Centre of Excellence 184

AI Governance Centre of Excellence Strategies 186

CoE Management Structure 190

Key AI Governance Roles and Responsibilities 192

Structuring My Board of Directors 202

AI Governance Oversight Committee Recommendations 204

vii

TABLE OF CONTENTS

Summary and Thoughts 205
Quiz 206

Chapter 8: Managing and Addressing AI Compliance 209

The AI Compliance Hype and History 209
Proposed, Enacted, and Emerging AI Regulations 210
Individual States Begin to Regulate AI in Absence of Federal Legislation 212
AI Compliance: The Evolving Landscape 217
Practical Approach to Achieve AI Compliance 219
Preventing AI Compliance Surprises and "Gotchas" 222
Balancing Innovation and AI Compliance 224
Summary and Thoughts 226
Quiz 227

Chapter 9: Integrating AI Governance with Enterprise Governance Risk and Compliance 231

What Is Enterprise Governance Risk and Compliance (EGRC) 232
Integrating AI Governance and Risk Management with the Broader Enterprise Governance, Risk, and Compliance (GRC) Framework 233
Integrating AI Risk Management with Enterprise Risk Management and eGRC 239
Auditing AI Controls and Processes to Validate Compliance 240
Summary and Thoughts 245
Quiz 245

Chapter 10: AI Policy Management and Enforcement 249

What Is AI Policy Management 249
The AI Policy Lifecycle 252
Integrating AI Policy Management with Key Business Policies 255

viii

Summary and Thoughts 256

Quiz 257

Chapter 11: Maintaining Privacy Within Your AI Governance Model 259

AI Actors Collaborating with Legal Counsel on AI Privacy Matters 262

AI Actors and Legal Counsel Collaboration 262

Summary and Thoughts 264

Quiz 264

Chapter 12: Human Oversight of AI Systems 267

The Inherent Limitations of AI 267

Human Oversight Challenges and Risk Mitigation 269

The Inherent Limitations of AI 270

Human Oversight Challenges and Risk Mitigation 272

Summary and Thoughts 274

Quiz 275

Chapter 13: The Power of Stakeholder Engagement in AI Governance 277

Strategies for Inclusive and Effective Decision-Making 277

Building Public Trust and Involvement in AI 282

Breakdown of Common AI Myths and Strategies to Debunk Them 283

Addressing the AI Skills Gap and Diversity Crisis 285

Systemic Barriers to Minority AI Entrepreneurs 286

Tailoring the Approach of Changing AI Myths for Diverse Cultural and Racial Groups 288

Summary and Thoughts 290

Quiz 291

TABLE OF CONTENTS

Chapter 14: Considering the Environmental Impacts of AI Systems 295

Fostering a Culture of Continuous Improvement 297

The Power of Algorithmic Efficiency 298

Knowledge Distillation and Quantization 299

Understanding the Full AI System Lifecycle 301

The Importance of Transparency...302

Integrating Sustainability into AI Design 303

Summary and Thoughts...304

Quiz...304

Chapter 15: Developing the Protocols for Rapid Response in Case of an AI-Related Crisis 307

Anticipatory Risk Assessment and Management............................308

Emergency Shutdown Procedures 309

Communication Transparency...310

Collaborative Incident Response...311

Forensic Analysis and Post-crisis Review.......................................312

Ethical Considerations in AI Crisis Management............................313

Continual Evolution and Vigilance 316

Summary and Thoughts...318

Quiz...319

Chapter 16: Capacity Building for AI Actors 321

Developing Skills and Knowledge Among AI Actors and Stakeholders.............321

Well-Governed AI Design for Developers 322

AI Literacy for End Users..323

AI Literacy for Business Executives 325

Convincing My Board of Directors and Senior Leadership Team to Use AI 327

Summary and Thoughts 328

Quiz 330

Chapter 17: Intellectual Property Rights with AI Technologies 333

Addressing the Challenges Related to the Ownership and Sharing of
AI-generated Content and the Technologies Themselves 334

Human Authorship vs. Machine Agency 334

Commercialization Concerns 336

Legal Frameworks and International Collaboration Can Help with
AI IP Rights 337

Global Standards and Norms for AI Governance and
Model Risk Management 338

The Power of Shared Governance 339

Summary and Thoughts 340

Quiz 341

Chapter 18: Auditing AI Systems 343

Traditional IT Audit Programs vs. AI Audit Programs 344

AI Auditor Essential Skillsets 345

AI Audit Program and Universe 345

Auditing Process, Risk, and Controls 348

Why the Audit Committee Needs to Be Involved 349

Defining AI Audit and Risk Assessment Cycles 350

Conducting an AI Readiness Assessment 352

Addressing Exceptions, Findings, and Remediation 357

Summary and Thoughts 361

Quiz 362

TABLE OF CONTENTS

Chapter 19: AI Model Inventory and Facts 365
Model Identification and Factsheet...366
Considerations When Choosing a Discovery Tool............................368
Summary and Thoughts 370
Quiz 371

Chapter 20: AI and Enterprise Architecture...................................373
The Strategic Union: Unveiling the Empowering Role of Enterprise Architecture in AI Initiatives 373
The Amalgamation of AI, EA, and Business Strategies....................374
Designing the AI Landscape with Enterprise Architecture 376
Enterprise Architecture during AI Implementation and Ongoing Management...377
The Benefits of Amalgamating AI, EA, and Business Strategy Teams 379
Summary and Thoughts 380
Quiz 381

Chapter 21: Aligning AI Governance, AI Development Lifecycle, and Systems Development Lifecycle Processes..............................383
Technology Obsolescence and Risks 384
Defense-in-Depth Strategies Within the SDLC and AI Development Processes...387
Rethinking the Development Lifecycle: Building AI in a New Age....388
The Symbolic Dance Between DataOps and ModelOps 390
Red-Teaming and Offensive Security Testing During the AI Development Lifecycle Process 394
Offensive Security Testing Strategies for AI Systems 395
Aligning AI Development Lifecycle and SDLC – Case Study............398
AI Development Lifecycle to Traditional SDLC Lessons Learned.....402

Getting the SDLC Process Ready for the AI Development Lifecycle Process 403

Summary and Thoughts 405

Quiz 406

Chapter 22: AI Through the Lens of Non-technical Business Leaders: Embracing AI with Caution 409

Understanding the Precautious Adoption of AI Systems from a Business Perspective 410

Understanding the Consumer's Take on AI in Their Daily Lives 415

Summary and Thoughts 415

Quiz 416

Chapter 23: Navigating the AI Frontier with This Sales Bible: Sales and Marketing Strategies for AI Governance and Risk Management Solutions 419

Understanding the AI Governance and Risk Management Landscape from a Sales and Marketing Viewpoint 420

Crafting a Compelling Value Proposition 423

Developing Targeted Sales and Marketing Strategies 424

Sales Techniques for AI Governance and Risk Management 428

Measuring and Optimizing Sales and Marketing Performance 436

Summary and Thoughts 441

Quiz 442

Glossary 445

Index 465

About the Author

With over twenty-five years in senior leadership, providing consultations and teams with forward-looking solutions, **James Sayles** possesses the AI Strategy and Governance Executive and CISO certification and executive experience necessary to carry out senior leadership responsibilities effectively. He has a Big 4 tenure, and using his Governance Risk and Compliance (GRC) strategy experience, he has executed intelligent automation and cybersecurity strategies for Fortune 10, Fortune 100, and mid-market companies across industry diversification in the public and private sectors. He's passionate about leading the charge for ethical and secure AI systems and has helped business executives with the necessary disciplines, thought leadership, and managerial responsibilities defined as good governance. He's in the industry to drive long-term, sustainable success and growth for organizations.

In addition, James is an entrepreneur and loves solving complex business issues. His career exhibits the ability to solve business problems at the intersection of technology risks and compliance validation. He founded and co-founded a few companies specializing in AI development, risk-governance strategies, enterprise governance, risk and compliance ecosystems, and cybersecurity centers of excellence (CCoE). Artificial intelligence governance and risk management is a current focus and a charge he is pioneering, under the alias of "The Sensei of AI Governance and Risk Management." Additionally, he continues to seek ways to automate the GRC – governance, risk, and compliance process workflows using AI. He recently consulted with a major healthcare and insurance and financial services client to automate their regulatory change management and compliance validation processes. He often has deep business-to-IT strategy expertise within the board and senior leadership advisory arena,

ABOUT THE AUTHOR

taking on advisory roles. He has unique financial services, government, and healthcare expertise. He is a recognized thought leader and driver of AI governance and secure, compliant, and risk-intelligent methods for major corporations.

Preface

Artificial intelligence (AI) is changing how we manage technology and business operations while providing the benefits of automation. AI offers several business advantages; however, the risks associated with identifying, assessing, and managing AI models are lagging. Moreover, the haste in developing AI models comes with its share of ethical concerns. AI governance and model risk management are not options or choices; they are the conduits to ensure transparent, unbiased, and trustworthy AI systems. This book provides practitioners, business executives, and boards of directors with a principled performance model and the solutions and strategies needed to design, validate, and deploy AI governance and model risk management control procedures. This book will ensure your AI governance and model risk management frameworks are operationally effective and lead you toward responsible intelligent automation.

AI governance and model risk management require proper planning, strategy, and executive oversight. While this may be overwhelming for small to mid-size organizations, this book of guiding principles will hopefully simplify the process, providing prescriptive methods for business leaders, MLOps, developers, and stakeholders. This book prepares AI actors and practitioners with the strategies and guidance to identify, assess, and manage the risks associated with artificial intelligence models. Furthermore, by providing the reader with practical examples and use cases, the reader will gain the essential knowledge to implement responsible and ethical AI models and systems that align to an organization's business goals and intelligent automation objectives.

PREFACE

What This Book Covers

AI strategy and governance is much more than technology alone. It is about people, processes, and technology, converged with risk intelligence, accountability, and responsibility. This book emphasizes the importance of cross-team collaboration, communication, and continuous improvement in building trustworthy AI systems. It provides practical use cases and methods to control and mitigate risks related to bias, inaccuracies, data quality, model drift, and hallucinations. Whether you're a technical expert, an ethics officer, or corporate counsel, a business leader, a consumer, or a person selling AI solutions, this book will empower you to play an active role in shaping the future of AI systems.

Overview of Chapters

Chapter 1, The Current State of AI Governance and Model Risk Management – This chapter explores the current landscape of AI governance and model risk management, highlighting organizations' challenges and inconsistencies in establishing effective frameworks. This chapter reinforces the need for responsible oversight, ethical considerations, and risk mitigation strategies as AI systems become increasingly integrated into decision-making processes across various industries.

Chapter 2, Aligning AI Strategy and AI Governance – This chapter emphasizes the critical relationship between AI strategy and AI governance. It affirms that governance frameworks must complement a well-defined AI strategy to ensure responsible and ethical AI implementation, mitigate risks, and maximize the benefits of AI technologies.

Chapter 3, How to Sound like an AI Governance and Model Risk Management Guru – Chapter 3 provides a comprehensive overview of AI governance and model risk management, including the definition of key terms and concepts. It discusses the drivers behind the growing importance of these disciplines, including the increasing power and complexity of AI systems, high-profile incidents, emerging regulations, and the recognition of potential risks.

Chapter 4, Designing a Well-Governed AI Lifecycle Model – From model proposal and feasibility assessment to ongoing monitoring and continuous improvement, Chapter 4 outlines the stages of a well-governed AI lifecycle model. It emphasizes integrating governance considerations at every stage to ensure responsible and ethical AI development and deployment.

Chapter 5, Aligning AI Governance with Other Internal Governance Models for Trustworthy AI: "The Convergence of Governance Frameworks" – This chapter explores the integration of AI governance with other internal governance models within an organization. It emphasizes aligning AI governance with data governance, IT governance, cybersecurity governance, and model risk management to create a comprehensive and effective framework for trustworthy AI.

Chapter 6, Designing Your AI Governance Framework – This chapter provides a practical guide for designing an AI governance framework tailored to an organization's specific needs and risk profile. It includes the key components of such a framework, including ethical principles, policies, processes, roles and responsibilities, and technology tools.

Chapter 7, AI Governance and Oversight Model – Chapter 7 provides the methods in establishing and AI Governance Center of Excellence (CoE) and its role in overseeing organizational AI initiatives. It discusses the CoE's management structure, key roles and responsibilities, and strategies for effective oversight, ensuring responsible and ethical AI development and deployment.

PREFACE

Chapter 8, Managing and Addressing AI Compliance – This chapter explores the evolving landscape of AI compliance, including the emergence of regulations and standards worldwide. It provides a practical approach to achieving AI compliance, emphasizing the importance of proactive measures, risk assessment, and continuous monitoring to ensure AI systems adhere to legal and ethical requirements.

Chapter 9, Integrating AI Governance with Enterprise Governance Risk and Compliance – Chapter 9 discusses integrating AI governance with broader enterprise governance, risk, and compliance (eGRC) frameworks. This chapter reinforces the importance of aligning AI initiatives with existing risk assessment protocols, compliance standards, and ethical guidelines to create a holistic approach to managing AI-related risks.

Chapter 10, AI Policy Management and Enforcement – This chapter explores the importance of AI policy management and enforcement within an organization. It outlines the AI policy lifecycle, from development and dissemination to implementation, monitoring, and continuous improvement. Additionally, it addresses the challenges and best practices in ensuring adherence to AI policies.

Chapter 11, Maintaining Privacy Within Your AI Governance Model – This chapter explores the importance of prioritizing privacy in AI governance models, including embedding privacy by design principles, enforcing strict data governance, and implementing privacy-enhancing technologies to protect sensitive information throughout the AI lifecycle.

Chapter 12, Human Oversight of AI Systems – This chapter explores the concept of human oversight in AI systems, emphasizing the need for human judgment and intervention in critical decision-making processes. Learn about AI's limitations, the importance of explainability, and strategies for ensuring accountability and building trust in AI systems.

Chapter 13, The Power of Stakeholder Engagement in AI Governance – Explore the significance of stakeholder engagement in AI governance, including strategies for inclusive decision-making, building public trust, addressing the AI skills gap, and tailoring approaches to debunk common AI myths for diverse cultural and racial groups.

Chapter 14, Considering the Environmental Impacts of AI Systems – Chapter 14 examines the environmental impact of AI systems, focusing on energy consumption, hardware waste, and the potential for AI to contribute to sustainability efforts. It emphasizes the importance of transparency, measurement, and responsible practices in mitigating AI's ecological footprint.

Chapter 15, Developing the Protocols for Rapid Response in an AI-Related Crisis – This chapter outlines the importance of proactive AI crisis management and discusses anticipatory risk assessment, developing rapid response protocols, emergency shutdown procedures, transparent communication, collaborative incident response, forensic analysis, and ethical considerations in managing AI-related crises.

Chapter 16, Capacity Building for AI Actors – This chapter focuses on developing skills and knowledge among AI actors and stakeholders and discusses training programs for regulators, developers, and end users. It emphasizes the importance of AI literacy, ethical AI design, bias mitigation, and explainability.

Chapter 17, Intellectual Property Rights with AI Technologies – Addressing the challenges of owning and sharing AI-generated content and technologies, this chapter explores the complexities of human authorship vs. machine agency, commercialization concerns, and the need for legal frameworks and international collaboration to address intellectual property rights in AI.

Chapter 18, Auditing AI Systems – With a comprehensive guide on auditing AI systems to ensure compliance and risk management, Chapter 18 discusses the differences between traditional IT and AI audits and outlines key focus areas, including a sample audit program. It emphasizes the importance of involving the audit committee and addressing exceptions and remediation.

Chapter 19, AI Model Inventory and Facts – With an emphasis on the importance of maintaining an AI model inventory and accompanying AI factsheets, this chapter outlines the process of model identification, information collection, and factsheet creation. It also addresses the challenges of maintaining an updated inventory and strategies to overcome them.

Chapter 20, AI and Enterprise Architecture – Exploring the strategic union between AI and enterprise architecture (EA), Chapter 20 discusses how EA can empower AI initiatives by providing a holistic view of the organization's technological landscape, facilitating integration, ensuring scalability, and addressing governance and risk management concerns.

Chapter 21, Aligning AI Governance, AI Development Lifecycle, and Systems Development Lifecycle Processes – Chapter 21 explores aligning AI governance with the AI development lifecycle (AIDLC) and the traditional systems development lifecycle (SDLC). It looks at the challenges of technology obsolescence, emphasizes defense-in-depth strategies, and provides a case study illustrating the integration of AIDLC and SDLC.

Chapter 22, AI Through the Lens of Non-technical Business Leaders: Embracing AI with Caution – With a deep dive into the perspective of non-technical business leaders on AI adoption, Chapter 22 discusses concerns about costs, workforce impact, and ethical implications, acknowledging the potential benefits of AI in driving innovation and efficiency. It emphasizes the importance of building AI literacy among executives and adopting a cautious yet proactive approach to AI implementation.

PREFACE

Chapter 23, Navigating the AI Frontier with This Sales Bible: Sales and Marketing Strategies for AI Governance and Risk Management Solutions – Offering a close look at sales and marketing strategies for AI governance and risk management solutions, this chapter provides insights into the market landscape, crafting a compelling value proposition, developing targeted strategies, and measuring and optimizing sales and marketing performance. It emphasizes the importance of building trust, educating customers, and demonstrating the ROI of AI governance solutions.

AI Model Aliases

Table 1. Table of Aliases

Item	Also Known As
AI-G	AI Governance
MRM and MRG	Model Risk Management and Model Risk Governance
Model Facts	AI Facts and Inventory Management
Resource	Application

Key Governance Concepts and Principles

Following is a comprehensive list of common AI governance and model risk management terms, including governing processes, guiding principles, and technologies I found helpful in implementing comprehensive AI governance centers of excellence and frameworks.

Core Governance Concepts

- **AI governance is the basis for a responsible and transparent AI system.** At its core, it provides your organization with the essential policies, structure, and roles to develop trustworthy and accountable AI systems.

- **AI ethics is a system of moral principles and techniques** intended to promote the safe development and responsible use of AI-infused technologies.

- AI governance establishes oversight, accountability, ownership, and responsibility for AI systems.

- **Organizations must develop AI systems using a centralized governance structure and the essential controls and processes to manage AI risks.**

- **AI developers and MLOps personnel must include metrics and KPIs** throughout the model's development, validation, and deployment phases.

- **The AI governance strategy must assure model explainability,** how AI systems perform decision-making, and the data used to train models.

- **A strategic AI governance framework must have operational components to validate transparency,** and its logic should always be monitored.

- **AI fairness ensures AI models do not disseminate biased and sensitive demographics** such as race, gender, socioeconomics, or other personal specifics.

Data Governance

- **Data governance is essential to AI model training**, ensuring the accuracy, completeness, consistency, timeliness, and relevance of data used for AI development.

- **Data governance is essential to the model training process.** It ensures data integrity and AI trustworthiness and prevents unauthorized access and use for model development.

- **Data governance reduces the risk of biased AI outcomes**, unfair decision-making, and inaccurate datasets.

- **Data lineage tracks the source of training data** and usage throughout the AI development lifecycle.

- **Metadata helps algorithms sort through massive datasets** while exposing data attributes, i.e., KPIs and metrics related to data quality and accuracy, patterns, ownership, and origin.

- Data security controls like encryption, prohibitive access, and backups **will prevent unauthorized access and data exfiltration and accelerate recovery from potential ransomware attacks**.

- When collecting and using data for AI systems and to train models, **privacy by design is essential to protecting personally identifiable information (PII)**.

Model Governance and Risk Management

- Model governance ensures that high-risk AI systems and algorithmic outcomes and decisions can be explained and substantiated.

- As part of the AI governance and model risk process, model validators must be well-trained in testing procedures to evaluate accuracy, performance, and fairness under different conditions.

- AI models and development activities must always be tracked and auditable. The ability to have AI facts, inclusive of versions and data lineage, must be evident in the governance process

- Model versioning is a key process that enables AI developers and other relevant AI actors to restore previous model versions in a timely manner.

- Model risk management strategies must include control activities, policies, and procedures that formalize AI model risk management activities for deployment. Model governance identifies and mitigates potential risks associated with AI systems, including biases, adversarial AI attacks, and model hallucinations.

Operational Governance

- Standards for controlled deployment of AI models and AI-based risks must be identified before promoting AI models into production.

- Integrating human judgment, a.k.a. Human-in-the-Loop (HITL) and oversight into AI decision-making processes, especially for high-risk or highly sensitive AI models and systems, improves AI trust.

- Organizations must establish feedback procedures for relevant AI actors and stakeholders regarding AI model outputs, enabling continuous improvement and bias mitigation.
- Ongoing Monitoring: Tracking the performance, fairness, and robustness of AI models in production **to detect concept drift, biases, and unexpected outcomes.**

Core Principles

- Model validators, data scientists, and AI developers must diligently ensure AI governance and controls are in place to mitigate risks of unplanned results in development and production environments.
- Organizations should provide reasonable assurance of AI systems' resilience against errors caused by data inferiority, adversarial attacks, and model drift.
- AI systems must be safe and minimize the risk of harm they cause.
- Ensure that AI controls reasonably ensure that AI systems and data are protected from unauthorized access or attacks.
- Organizations must ensure the models include transparency, the use of data is minimized, and decisions can be justified.

Governance Frameworks and Standards

- NIST AI Risk Management Framework: A voluntary framework for managing AI risks
- IEEE Ethically Aligned Design (EAD): Principles for embedding ethics into AI systems

- OECD AI Principles: High-level policy guidelines for trustworthy AI
- EU's Proposed AI Act: Regulatory framework with risk-based classification of AI systems
- ISO Standards: Emerging standards for various aspects of AI governance

Roles and Responsibilities

- **AI Governance or Steering Committee** oversees AI strategy and risk management.
- **The Data Scientists** are responsible for model development, bias mitigation, and explainability.
- **An AI Ethicist** guides ethical considerations for AI development and use.
- **The Risk and Compliance Officer** ensures adherence to regulations and ethical standards.
- Business Stakeholders collaborate on defining AI use cases and governance requirements.

Tools and Techniques

- Bias Detection Tool is software for identifying biases in datasets and models.
- Model Interpretability Methods – techniques like LIME and SHAP for understanding your AI model logic.
- Adversarial Robustness Testing: Evaluating AI resilience to malicious attacks.

- Differential Privacy: Techniques for anonymizing data to protect privacy.
- AI Auditing Frameworks encompass structured processes for assessing AI processes, risks, and controls with governance principles.

Additional Considerations

- Human-in-the-Loop: Involving human oversight in critical AI decisions
- Social Impact: Assessing the potential societal consequences of AI systems
- Continual Monitoring: Ongoing assessment of AI performance in real-world environments
- Incident Response: Processes for managing AI failures or ethical breaches

Acknowledgments

I want to express my sincere gratitude to Ramesh Balanagu for his invaluable contributions and expert guidance on AI strategies and enterprise architecture. His deep knowledge and insights, so generously shared, were truly instrumental in shaping my understanding of this complex and rapidly evolving field. Ramesh's expertise proved timely and essential, providing the critical information necessary for the successful development and implementation of AI systems within corporate America. I especially appreciate his ability to explain complex concepts clearly and conclusively, making them readily accessible and actionable. His real-world comments were invaluable in bridging the gap between theory and practice.

Thank you, Ramesh, for your dedication to sharing your knowledge and significantly impacting this book.

CHAPTER 1

The Current State of AI Governance and Model Risk Management

In the fast-paced domain of artificial intelligence (AI), governance, oversight, and model risk management are center stage and at the top of minds for business executives. The present state of AI governance and risk management is undergoing development, but it is also one of inconsistency and growing pains. As organizations continue integrating AI systems into their operations to automate decisions and business processes, the need for responsible oversight has never been more important. In fact, organizations across industries are now starting to grapple with the complexities of establishing AI governance frameworks. Many face the challenges of pulling a community of stakeholders together to implement even the foundational components of an AI governance model that aligns with their AI operations. AI governance frameworks are the go-to source and strategy to ensure their AI systems are developed and deployed ethically, transparently, and with a clear understanding of the potential risks. However, AI governance and model risk management are not new, especially within the financial services industries (FSIs). Many FSIs began to address AI governance and model risk management in the early 2010s, with substantial momentum within the last five years. In contrast, other industries have moved much slower over the same five years, taking a "crawl" approach as they come up to speed in adopting and instituting cross-functional teams dedicated to AI ethics and transparency. We are seeing more organizations employ governance-savvy data scientists, add internal and outside legal counsel, experts, and AI ethicists to their teams to navigate the landscape of explainability, bias prevention, risk mitigation, and positive societal impact.

Model risk management, a discipline traditionally adopted by financial institutions, is now finding new relevance in energy, manufacturing, and many technology sectors. AI's dynamic and sophisticated nature necessitates thorough validation and approval processes, continuous monitoring for performance degradation, and strategies to mitigate risks associated with inaccurate data sources and biased outcomes. While realizing these necessities, still, there are only a few tools and platforms on the market to provide a reasonable assurance for governance oversight and internal control procedures. Many Governance, Risk, and Compliance (GRC) solution providers are now re-tooling their solutions for this necessary AI process.

Despite the growing recognition of the importance of AI governance and model risk management, a consistent and universal framework is lacking. While it is impractical to expect a one-size-fits-all solution, a common set of control objectives and governance framework across industries and geographies should be sought. One must also expect different industries and applications to have tailored governance and risk management approaches. For instance, healthcare organizations may prioritize patient safety and data privacy, while FSIs may focus on financial stability and regulatory compliance. Nonetheless, there's a growing consensus on the need for clear accountability, transparency, and human oversight in this sophisticated technology called intelligent automation.

CHAPTER 1   THE CURRENT STATE OF AI GOVERNANCE AND MODEL RISK MANAGEMENT

Fast-forward to 2024, a common theme with customers is deploying transparent and compliant AI systems, while ensuring these systems provide business efficiencies and competitive advantages, which is ascertained through proper governance and risk management strategies. The framework components they seek include establishing metrics and key performance indicators, ethical guidelines for development, conducting thorough risk assessments, defining accountability structures, and ensuring that AI systems are explainable and auditable. Collaboration within think tanks and regulatory bodies is also becoming more common, aiming to establish best practices and standards for AI governance on a broader scale. The journey toward effective AI governance is ongoing, but the general commitment to building responsible and trustworthy AI applications is a shared goal across industries and global regions. Here are a few specific callouts as to the current state of this topic:

- **Rising Awareness** – The relative risks with artificial intelligence applications and the importance of governance standards are receiving greater attention these days. The effects of incidents and proposed regulations are two factors driving attention and concern on a global scale, specifically in the news and the World Leaders' Policy Commission.

- **The Need for AI Governance and Oversight Structures** – AI governance and risk management frameworks are what fuel sound AI strategies. Additionally, the need for a global oversight body or consortium driving a common regulatory framework will ensure global standards and objectives are applied, with minor variations based on industry or applications, for instance, the AI Act of the European Union and standards organizations like NIST or private entities. The world of AI currently does not have any real sign of which, if any, of these entities is going to recommend the most effective governance objectives or regulatory framework in the long run. Of course, you can imagine that various governmental entities might have different conceptions. In the meantime, the industry is ripe for framework development and standards for AI governance and model risk management.

- **AI Governance and Risk Management Maturity Gaps** – A disconnected world of standards and regulations yields a system of compliance battles and chaos. However, with the assistance of industry experts, a few global trends are starting to emerge, but the maturity levels remain low, and gaps are wide.

- **Major Healthcare Companies and Financial Services Organizations** in highly regulated industries are slightly ahead of other industries and many small businesses. Mid-size organizations, despite having footprints of AI systems, are struggling to find necessary capital to develop AI governance and oversight structures and frameworks.

- **AI Technology Challenges** must be addressed and resolved to ensure AI is fair, unbiased, explainable, and continuously monitored. Another issue that must be addressed is the inability of a single human to manage and comprehend large, complex AI models. When these large, complex models keep learning from new data fed to them, they can become unreliable and dangerous.

CHAPTER 1   THE CURRENT STATE OF AI GOVERNANCE AND MODEL RISK MANAGEMENT

- **Skills Shortage in AI Governance** – There is a shortage of professionals with artificial intelligence, governance, and risk management skills.

- **Practical AI Development and Deployment** are important shifts in the workforce today. The conversations are moving from theoretical principles to actual, useful advice and technologies with real-world applications.

- **Focus on Fairness and Bias Risks** – Our top priority is to develop effective AI governance strategies that can mitigate fairness and bias risks – a primary reason this book is being written today. But that raises a myriad of questions that we also must address and answer. For example, what is fairness relative to AI? And if it's easy enough to understand, can we actually achieve fairness?

- **Model Explainability Is Necessary** – Making AI decision-making models transparent and understandable to stakeholders and data subjects is another top priority for industry thought leaders.

- **The convergence of model risk management (MRM) and AI governance is essential, and guiding principles must be developed to address** significant AI challenges. It's imperative for these disciplinarians to align on control procedures to ensure that AI models are sufficiently managed in a triage-ready manner. And no secret here; in MRM itself, the appearance of these collaborators – between AI ethics and MRM – is otherwise seen as inherently good.

5

- **Oversight and Regulations** – AI-specific regulations will likely increase in the next few years, and organizations must comply with them, rationalizing the development of AI governance and risk strategies.

Here is a common question I get. "What are some of the obstacles you are seeing, and what can organizations do to get moving?"

Companies are struggling with the following:

- **Balancing Governance and the Lack of Government Oversight** – Artificial intelligence is a rapidly advancing technology that is hard for regulatory agencies to develop and achieve oversight, requirements validation, and governance.

- **Balancing Risks and AI Innovations** – Finding the right equilibrium between innovation and the safeguards against risks – to encourage innovation is currently a challenge with many precedents to look to but a governing conundrum for organizations.

- **Instituting AI Ethics** – It's not easy to turn high-minded principles into day-to-day realities. Translating ethical principles for AI development can be challenging and time-consuming.

- **Worldwide coordination of AI control standards** is a global issue that affects multinational firms. International alignment and collaboration are still lacking or inconsistent at best.

CHAPTER 1   THE CURRENT STATE OF AI GOVERNANCE AND MODEL RISK MANAGEMENT

Additional Challenges:

> **AI Governance Structure and Strategy** – Most companies have decentralized AI development and governance operations. For example, multiple lines of businesses within an organization have separate data scientists, MLOps Engineers, and AI developers, all with unique control objectives. While this may work for some organizations, best practices have proven that a more centralized model is cost-effective and will drive enterprise-wide compliance.

Organizations can use a few options to accomplish a more centralized governance model.

Establish a Centralized Governance Body

- **Option 1: AI Center of Excellence** – Establish an AI Center of Excellence to develop and enforce AI governance standards. This would be a team of individuals with a range of specific backgrounds (technical, legal, compliance, and business) who would establish these standards and then ensure that a more centralized governing body and approach to risks and controls are implemented.

- **Option 2: AI Governance Committee** – Create an AI governance committee of senior leaders across relevant business units who would review high-risk AI projects, set overarching strategic direction, and resolve conflicts that arise between different departments.

- **Hybrid Approach** – A Center of Excellence can collaborate with a governance committee to execute individual business unit operations and provide strategic oversight and management of AI operations.

CHAPTER 1   THE CURRENT STATE OF AI GOVERNANCE AND MODEL RISK MANAGEMENT

The significance of AI governance and model risk management is increasing. While progress has been made, a globally unified and mature approach is still lacking. It is essential to anticipate future regulations and be prepared to develop internal expertise to adapt to these changes. Furthermore, our existing governance frameworks must continue to evolve and adapt to the shifting landscape.

Current Maturity Levels on AI Governance and Model Risk Management

The maturity of AI governance and model risk management varies significantly across industries, reflecting differing levels of AI adoption and regulatory pressures. For example:

> *Financial Services*: As early adopters of AI, financial institutions are often considered more mature in model risk management due to existing regulatory frameworks like SR 11-7 for model risk. However, the governance of newer AI technologies like deep learning is still evolving. Leading banks and financial services industries (FSI) are establishing AI ethics boards and investing heavily in tools and platforms for explainable AI (XAI), model facts and inventory, assessing and managing model risks, cybersecurity, and auditing to balance innovation with regulatory compliance and customer trust.

Healthcare: The healthcare sector is catching up quickly due to the risk of clinical failures and health risks and the sensitivity of data being used to train their model. While model risk management is crucial due to the potential impact of AI on patient outcomes, ethical considerations surrounding data privacy, algorithmic bias, and informed consent are paramount. Many healthcare providers are developing AI governance frameworks prioritizing transparency, fairness, and human oversight in clinical decision support systems.

Tech Industry: I often interact with major application and software solution providers, and 99.9% of them claim to have some machine learning or natural language processing algorithms within their solutions. What's alarming is that the tech giants at the forefront of "AI development" are lagging with their internal AI governance initiatives, including ethics boards and responsible AI principles. However, external scrutiny of their practices, specifically from the United States and the European Union, calls for stronger regulation of companies developing AI-based solutions, particularly regarding bias, misinformation, and the potential to misuse AI technologies.

Other Industries: Industries like manufacturing, retail, and energy are gradually incorporating AI into their operations. While these industries move much slower, AI governance's maturity levels vary widely. Some leaders proactively establish frameworks, while others are still in the early stages of recognizing and addressing AI-related risks. Regulatory frameworks specific to AI are less developed in these sectors, but growing awareness and best practice sharing are driving progress.

The current state of AI governance and model risk management is rapidly evolving. While some industries, like finance, are relatively mature due to existing regulations, others rapidly develop frameworks and practices to ensure responsible AI adoption. The challenge for all organizations is staying ahead of the curve and balancing innovation with ethical considerations and risk management practices.

Assessment of the Proficiency Levels of AI Governance Strategy Experts

To navigate the complex landscape of AI governance and model risk management, a diverse skill set is paramount, yet the skills in the market required to drive a comprehensive AI governance and model risk management strategy are limited. A deep understanding of AI and machine learning technologies is given, encompassing the cross-domain knowledge of algorithms, data governance, cybersecurity, controls assessment, and model evaluation techniques. But equally important are "soft" skills such as analytical thinking, policy and control development, ethical reasoning, and strong communication to bridge the gap between technical and non-technical stakeholders. Additionally, knowledge of regulatory landscapes and risk management frameworks is vital for ensuring compliance and building trust. Individuals who can effectively implement AI governance and model risk management must possess a unique blend of technical, analytical, and interpersonal skills. Strong problem-solving and decision-making abilities are essential for identifying potential risks and developing appropriate mitigation strategies. Communicating technical concepts to a non-technical audience fosters transparency and builds trust in AI systems. Moreover, a detailed understanding of ethical principles and societal implications is necessary to ensure AI is used responsibly and for all to benefit. As AI adoption accelerates, the demand for these specialized skills is soaring. However, a notable shortage of skilled professionals in AI governance and model risk management persists. This talent gap poses a significant challenge for organizations seeking to build ethical and trustworthy AI systems. Universities, educational institutions, and companies are beginning to bridge this gap through targeted training programs, certifications, and interdisciplinary collaborations.

While the talent pool grows, organizations must prioritize cultivating and attracting individuals with the necessary skills to navigate AI's ethical and risk-related complexities. Investing in talent development and fostering a culture of responsible AI innovation are essential steps toward ensuring that AI systems are designed and implemented with fairness, transparency, and accountability at their core.

What Are the Big Four and Other Strategic Consulting Firms Doing to Address the Gap in Skills?

By leveraging the expertise and resources of the Big Four and other strategic consulting firms, organizations can accelerate their AI adoption journeys while mitigating the risks associated with this transformative technology. Here is a short list of what these companies are doing to ensure the population of AI systems is ethical and trustworthy.

CHAPTER 1   THE CURRENT STATE OF AI GOVERNANCE AND MODEL RISK MANAGEMENT

CyberOne, LLC of Plano, Texas, is taking a thought leadership role in providing strategic AI Governance and Model Risk Management consulting services.

- **Centralized and Unified AI Governance Framework** – CyberOne has developed a multi-governance approach to its AI governance and model risk management framework. The framework focuses on AI strategy and AI governance alignment, multiple governance and risk management process integration, AI policy development and management, and regulatory compliance change management. The framework addresses AI transparency, AI explainability (XAI), AI Facts, privacy impact assessments, fairness, trustworthiness, responsibility and accountability, AI audit readiness, third-party (AI) management, feasibility, AI DevOps, validation and approval, deployment, and continuous monitoring. CyberOne also provides strategic consulting in developing AI governance oversight models and Centers of Excellence (CoE).

- **Cybersecurity and Data Governance** – CyberOne's AI governance and risk management solution includes publications for secure and safe AI systems and model accuracy using comprehensive data governance and data validation solutions, publishes thought leadership, and offers consulting services on AI governance and risk management.

- **Industry Coverage** – CyberOne provides AI governance and model risk management solutions for mid-size financial services, healthcare, energy, technology, and other regulated industries.

CHAPTER 1   THE CURRENT STATE OF AI GOVERNANCE AND MODEL RISK MANAGEMENT

Deloitte

- **Trustworthy AI Framework** – Deloitte has developed a framework for trustworthy AI, emphasizing six dimensions: fair and impartial, transparent and explainable, responsible and accountable, safe and secure, robust and reliable, and respectful of privacy.

- **AI Institute** – Deloitte's AI Institute conducts research, publishes thought leadership, and offers consulting services on AI governance and ethics.

- **Model Risk Management** – Deloitte provides model risk management solutions tailored to financial services and other regulated industries, focusing on model validation, governance, and risk assessment.

PwC

- **Responsible AI Toolkit** – PwC offers a Responsible AI Toolkit to help organizations assess and mitigate AI risks, including bias, fairness, and transparency.

- **AI Risk and Controls Framework** – PwC has developed a framework for managing AI risks and controls that are aligned with industry best practices and regulatory requirements.

- **AI Assurance Services** – PwC provides independent assurance services for AI systems, assessing their ethical and operational risks.

EY

- **AI Advisory Services** – EY offers a wide range of AI advisory services, including strategy development, risk assessment, governance design, and model validation.

- **Trusted AI Platform** – EY has developed a Trusted AI Platform to help organizations monitor and govern AI systems, ensuring fairness, transparency, and compliance.

- **AI Lab** – EY's AI Lab focuses on researching and developing AI solutions, including tools for ethical and responsible AI.

KPMG

- **AI in Control Framework** – KPMG has created an AI in Control framework to help organizations design and implement effective AI governance structures.

- **AI Risk Management Services** – KPMG provides model risk management and validation services, focusing on financial and regulated industries.

- **AI Center of Excellence** – KPMG's AI Center of Excellence brings together AI, data science, and ethics experts to develop and deploy responsible AI solutions.

Common Themes

Across the list of consulting solutions mentioned, there's a clear focus on developing frameworks, tools, and services to help organizations navigate the complex landscape of AI governance and model risk management. They also invest in research and education to build awareness and expertise.

Summary and Thoughts

The landscape and field of AI governance and model risk management are evolving rapidly. Businesses are beginning to recognize and we, as AI professionals, are stressing the urgent necessity of establishing controls

CHAPTER 1   THE CURRENT STATE OF AI GOVERNANCE AND MODEL RISK MANAGEMENT

and frameworks to mitigate the ethical, legal, and operational risks linked to AI systems. Although global regulations are not fully developed, many regions are introducing or considering laws tailored for AI, leading to increased emphasis on proactive compliance. It's imperative that organizations advance from basic checklists and incorporate responsible AI principles into their development procedures. Nonetheless, a notable disparity remains between acknowledging these risks and successfully implementing governance protocols and operational safeguards.

Organizations that fail to prioritize AI governance are exposed to significant legal, financial, and reputational hazards, such as penalties, legal actions, public confidence erosion, and reputation harm. Inadequate governance can result in ineffective AI development, erratic model performance, and heightened cybersecurity vulnerabilities. Moreover, disregarding AI governance obstructs innovation, poses competitive disadvantages, and neglects critical societal issues regarding the ethical use of AI.

Quiz

1. Identify three to four challenges organizations face when implementing AI systems.

2. What are the four major goals of AI governance and model risk management?

    A) Ethics

    B) Transparency

    C) Risk Intelligence

    D) Explainability

    E) Trustworthiness

CHAPTER 1   THE CURRENT STATE OF AI GOVERNANCE AND MODEL RISK MANAGEMENT

3. Which statement(s) is incorrect?

    A) It's not easy to turn high-minded principles into day-to-day realities and translating ethical principles for AI development can be challenging and time-consuming.

    B) The framework components organization must seek include establishing metrics and key performance indicators, ethical guidelines for development, conducting thorough risk assessments, defining accountability structures, and ensuring that AI systems must be audited at least annually.

    C) While the AI talent pool grows, organizations must prioritize cultivating and attracting individuals with the necessary skills to navigate AI's ethical and risk-related complexities.

    D) Worldwide coordination of AI control standards is a global issue that affects multinational firms.

    E) An AI strategy begins with a clear set of development standards and priorities to support the organization's mission and vision and defined in its strategic plans.

    F) A disconnected world of standards and regulations yields a system of compliance battles and chaos.

4. *True or False*: Financial Services and Technology Sectors are more mature from a Governance and Risk Management perspective

5. Identify the correct framework(s) currently available.

    A) The EU AI Act

    B) NIST AI RMF

    C) NIST CSF

    D) PCI-DSS

    E) CIPP EU/US

CHAPTER 1   THE CURRENT STATE OF AI GOVERNANCE AND MODEL RISK MANAGEMENT

6. Which of the following statement(s) is correct?

   A) Human-in-the-Loop involves human oversight in critical AI decisions.

   B) Organizations must prioritize cultivating and attracting individuals with the necessary skills to navigate AI's ethical and risk-related complexities.

   C) Organizations that centralize AI development and governance operations are more efficient in AI governance because more AI actors can ensure accountability.

   D) AI's purpose is to operate more effectively and control costs by reducing headcount.

   E) The United States is leading the charge in creating safe and trustworthy use of artificial intelligence.

7. Who is responsible for AI governance and risk management (select all that apply)

   A) Data Scientists and ML Engineers

   B) Business Unit Leaders and process owners

   C) Legal Counsel and Data Privacy Officers

   D) The Board of Directors and Chief Executive Officer

   E) All of the above

   F) None of the above

8. *True or False:* Organizations that prioritize AI Governance are more likely to succeed in business, gain competitive advantage, and reduce headcount.

CHAPTER 1   THE CURRENT STATE OF AI GOVERNANCE AND MODEL RISK MANAGEMENT

9. *Complete the following fact:* Organizations can address the gaps by ensuring the following:

   A) Balance governance and improve AI government oversight.

   B) Balance Risks and AI Innovations.

   C) Instituting AI Ethics.

   D) Worldwide coordination of AI control standards.

   E) Add your answer here.

10. What is the go-to source and strategy to ensure their AI systems are developed and deployed ethically, transparently, and with a clear understanding of the potential risks

    A) Cybersecurity controls and governance

    B) Enterprise Risk Management plan

    C) Internal audit of controls

    D) Global AI regulation and compliance program

    E) AI governance and model risk management

    F) A, C, D

    G) None of the above

CHAPTER 2

Aligning AI Strategy and AI Governance

An AI Strategy Without AI Governance Is an AI Failure

Why AI Governance Matters for AI Strategy and Vice Versa

Your AI strategy must outline and detail how your AI initiatives contribute to the goals and priorities of your organization. AI governance policies and standards must be in line with this strategy. A strategy helps to prioritize and identify the areas where AI can provide the most value, determining which projects and use cases require the most comprehensive governance procedures. A successful AI strategy assesses possible risks and ethical concerns related to different AI applications. This process guides the proactive steps taken within the governance framework. As such, the AI strategy validates resource allocation to AI governance by showcasing how it protects the organization and promotes responsible AI utilization in line with business goals.

How an organization uses AI should support its vision and goals. John Kamensky, Emeritus Senior Fellow and Director of the IBM Center for the Business of Government, said, "An AI strategy or initiative begins with a clear set of goals and priorities designed to support the organization's mission and vision and defined in its strategic plans."

CHAPTER 2 ALIGNING AI STRATEGY AND AI GOVERNANCE

How the organization sets out its AI strategy governs how it will use AI technology in line with core business objectives and explore available opportunities. In return, part of the AI strategy aims to communicate to stakeholders what AI is practically doing in each important business area. AI strategy differs from an IT or data strategy, but those strategies must be in place. Since 2003, industry analysts have been reporting that AI strategy is an instrument for the executive management team to use in executing AI's role in the business and that AI transforms business models even as it requires changes in IT and data infrastructures.

The Risks of an AI Strategy Without Governance

Without a well-defined governance framework, there is a potential risk of developing governance policies that fail to address the risks relevant to the most critical AI use cases or unnecessarily impede innovation, resulting in a lack of alignment. It becomes more probable that you will react to AI incidents after they occur rather than proactively working to prevent them, which could harm your reputation and diminish trust in your AI systems. Not having a well-defined plan for AI could result in overlooking or missing opportunities to capitalize on AI for competitive benefits and positive effects. Without strategic prioritization, there is a risk of investing in governance efforts that fail to attain significant reductions in AI risk while greatly negating business value generation, ultimately leading to the wastage of important resources, including human and financial capital. Even a basic AI strategy that evolves incrementally is better than none. Consider starting with these foundational elements:

- **High-Level Goals** – What problems are you addressing, what business advantages are you seeking to accomplish with AI systems, and how well does it align with your broader business vision and mission? The following hypothetical scenarios provide additional color and context to the question:

CHAPTER 2   ALIGNING AI STRATEGY AND AI GOVERNANCE

- A Healthcare Organization
  - **Problem** – Inefficient patient triage and diagnosis procedures in rural areas.
  - **Broader Business Mission** – To improve access to healthcare benefits and services for underserved populations.
  - **Proposed AI Solution** – Deploy an AI-powered diagnostic support tool for telehealth consultations.
- A Manufacturing Company
  - **Problem** – High operational costs and low productivity due to unplanned equipment downtime and outages.
  - **Broader Business Mission** – Drive operational efficiency and sustainability.
  - **AI Solution** – Predictive maintenance system to detect equipment failures before they occur.
- **Areas of Focus** – Which areas could benefit the most from AI, and where could it reasonably be used? (e.g., customer service, operational efficiency, decision-making)?
- **Key Areas for AI Value Creation**
  - Customer Service
    - **Intelligent Chatbots** – Provide 24/7 support, personalize interactions, and quickly resolve routine issues.

- **Sentiment Analysis** – Understand customer emotions and tailor communication for better experiences.
- **Recommendation Engines** – Offer tailored products or services based on past behavior and preferences.

- **Operational Efficiency**
  - **Process Automation** – Automate repetitive, rules-based tasks to free up employees for higher-value work.
  - **Predictive Analytics** – Forecast demand, optimize inventory levels, and improve supply chain efficiency.
  - **Anomaly Detection** – Identify unusual patterns in equipment data to prevent failures or optimize maintenance schedules.

- **Decision-Making**
  - **Data-Driven Insights** – Analyze complex datasets to reveal trends and patterns humans might miss.
  - **Risk Assessment** – More accurately evaluate and prevent healthcare waste, faster and more accurate processing of health insurance applications, claims, or financial transactions.
  - **Enhanced Diagnostics** – Support medical professionals in image analysis and disease detection.

Which areas could benefit the most from AI, and where could it reasonably be used? To identify areas where artificial intelligence can be most beneficial, first look for places where AI can be quickly deployed and will make an immediate impact with minimal effort and lower risk. This analysis will allow you to realize the benefits and demonstrate the value of AI technology, achieve accelerated wins, and build positive momentum within your organization.

Ethics. When it comes to the development and use of AI, what do you consider to be the most important ethical principles?

AI development and use should be actively accompanied by thoroughly examining fairness, privacy, and transparency issues and a solid understanding of accountability, ethics, and the potential social impact. For AI to be truly "good," its development and use must be socially responsible, meaning the AI system should add to a good society rather than decrease it, and the system must respect and comply with individual human rights. Adding this layer of human judgment by design ensures that your AI system and models can and will be trustworthy and ethical.

Because AI governance and AI strategy are entwining and mutually reinforcing, they must evolve together. This approach ensures continuous alignment of your AI strategy and business goals while maintaining ethics, transparency, and explainability. Otherwise, you will increase the risk of opaque and unintelligent automation at the expense of an effective AI strategy, core governance principles, and the ability to refine both as you gain more experience with your AI projects.

Foundation and Alignment

The first step is to evaluate your organization's AI capabilities. Conduct a thorough and wide-ranging assessment of your current state and document your findings. If you haven't already, now is a good time to seek tools to facilitate the evaluation of the current maturity of your AI capabilities, i.e., CMMI, NIST AI RMF, or COBIT and to estimate where on the AI hockey stick of potential returns, you may be headed. Once you know your current state, lay out the roadmap to how AI can and should develop within your organization in the near, medium, and long term.

CHAPTER 2   ALIGNING AI STRATEGY AND AI GOVERNANCE

Define Your Purpose and Objectives

Document the main business goals you're trying to address with AI. Ensure you clearly articulate business objectives to your stakeholders, aligning them to existing goals and objectives. Next, please focus on the set of problems you want AI to be able to solve and spell them out up-front. They should draw directly from your business strategy and be framed in your company's core ethical principles, which should be documented and understood. Your objectives should be mapped to what your organization aims to achieve and the strategy of how you plan to accomplish those objectives.

Align with the Regulatory Landscape

Stay abreast of the AI regulations and proposed acts for your industry or business sector. Proactively monitor for upcoming legal frameworks for AI, general (like the GDPR and CCPA) and others specific to AI, as you commence the AI model proposal and development phases. Implement control procedures and operational safeguards as part of your AI strategy. Be sure to incorporate key performance indicators and metrics that you can report to executive management and, ultimately, the Board, as they need to be aware of your automation approaches and business plans. Consider metrics for transparency, ethics, bias, privacy, and cybersecurity.

Identify AI Opportunities

Brainstorming Sessions

Generating ideas is sometimes best accomplished through a group dynamic. Business managers must carefully think about getting optimal results from all AI actors and stakeholders. Executives should foster an atmosphere and set the tone of the brainstorming session. This requires that executive leadership mandate trust and open and honest dialog amongst team members. These "think tanks" should feel comfortable in making suggestions and recommendations within reason. Employ cross-functional workshops that identify and pinpoint where things are going wrong and fail to meet intelligent automation objectives, where waste could be eliminated, and where the AI strategy could provide optimal business value to business units or processes. When organizations think about using and deploying AI, it often centers around reducing costs and resource constraints, making our business processes smarter, faster, or more efficient. How do we obtain competitive advantages using emerging or "bleeding" edge technologies? While these are good objectives, your AI strategy should include good governance with control procedures for your ability to trust in models, trust in data, and trust in processes.

Feasibility Assessment

Consider data availability, technical complexity, ethical implications, and potential ROI for each possible AI use case. Prioritize projects with demonstrable business value and a clear path toward implementation. Identify and examine all AI strategy business requirements, including how much data can be gathered and at what cost. Think about the complexity of the calculations. Consider the ethical implications for what can only be a pilot project (at least in the near term) and for any initiative. Take the various use cases, then rank them in terms of how favorable they may be. Also, consider the "whole-school experience." Have you seen anything like this before? Have you been able to observe calculations applied to massive datasets? If not, try to imagine it, then rank the use cases.

Prioritization Framework

To decide which projects to pay attention to, you need to start with standard criteria that all projects can be evaluated against. You cannot make this up as you go along; you must define your standard up-front during the decision stage. You need clear criteria for the same reasons that you need to have clear standards for success or failure; you need them because you're human, and you need to know why you made the decision you did.

Start with Pilot Projects

To gain experience, identify accelerators, seek quick wins and victories, and develop the energy to implement additional prescriptive AI capabilities. Start with small, clear, and realistic AI undertakings.

CHAPTER 2 ALIGNING AI STRATEGY AND AI GOVERNANCE

Building Your AI Capabilities

Skills Assessment

Start by identifying skills shortages and assessing your AI capabilities. Recognize the current lack of talent and resources required to carry out your AI strategy. Determine if the best way to address these opportunities or issues is to hire more experienced AI resources or retool the current resources based on business needs. Have secondary hiring plans if qualified AI resources aren't available or you cannot afford them. You can also consider the possibility of forging intellectual partnerships.

Technological Infrastructure

Assess your present requirements for storing data, compute (GPUs), and using AI governance platform solutions. Identify and document your technical requirements, e.g., cloud vs. on-premises, hybrid approaches, databases of data lakes, and ascertain whether your organization should have multiple environments or one.

Data Governance

Data governance is rapidly emerging as a new and important discipline for organizations. Data governance is how we manage data and access to make it more effective and secure. Data is the most important asset for decision-making and has many valuable uses outside of decision-making. This understanding has led many business organizations to establish data governance structures and frameworks to ensure data quality, security, and privacy. Don't be left out of this emerging trend and stay abreast of data governance best practices. Create effective data management systems and set up clear and comprehensive guidelines for all aspects of data handling – gathering, storing, accessing, and ensuring high quality. Establishing these systems is vital for ensuring AI systems' accuracy, reliability, and usefulness.

Experimentation Mindset

Foster an environment of innovation, experimentation, and learning where experimental AI model development is okay. In your "sandbox," you can and should give developers the autonomy and permission to innovate and try things they've yet to master while holding them to high ethical standards. This allows for rapid prototyping with human considerations at the forefront and a recognition that not all paths will lead to the same destination.

Governance, Communication, and Evolution

Establish AI Governance

Form an AI steering committee or working group with clear responsibilities for overseeing AI strategy. Set clear AI governance. AI shouldn't just be a fluffy buzzword; it requires the kind of structure and vision that compounds decisions and resources. Human decisions are still valid inputs to any AI model, and AI governance and accountability must be continuously improved. Chapter 7 provides more detail on this process.

Communication Plan

It's also important to overcommunicate the importance of proactive AI transparency and sharing AI initiatives and goals with AI stakeholders, employees, customers, and end users. Organizations that keep AI strategies transparent have a better chance of having them understood and better aligned with corporate goals and objectives.

Continuous Learning

The AI field is advancing quickly, so allocating and training AI resources is important to stay current with trends, research, and regulatory changes. Approach your AI strategy as a dynamic document that can adapt and enhance over time. Your plan should be flexible enough to pivot when necessary but well-governed to manage key risks and regulatory requirements.

Monitor and Measure

Establish measurable metrics and performance indicators for your AI projects: monitor return on investment, equity, and potential risks to inform future revisions and enhancements.

Refining Your AI Strategy for Principled Performance

Refining your AI strategy for principled performance is not merely a moral imperative; it's a strategic advantage that can differentiate your organization in the competitive landscape. By embracing and striving for AI principled performance, you build trust, mitigate risks, unlock the full potential of AI, and harness the unmistakable power of intelligent automation. By developing your AI strategy in alignment with business goals and continuously measuring and monitoring key performance indicators (KPIs), your AI system will increase its ability to be trustworthy while providing reasonable assurances of addressing business needs and procedures. Additionally, your AI system can seamlessly scale as your organization matures in its AI journey and embodies values like fairness, equity, and transparency. Transparently speaking, AI can increase your risk exposure and yield unintended consequences. Performing proactive inherent and residual risk assessments, soliciting stakeholder input, implementing strong data governance, emphasizing model explainability and well-defined accountability structures, and continuous learning is essential. Get as many eyes on your AI strategy as possible and push your developers to collaborate with your business stakeholders. Make accountable, risky decisions together, and once the AI system is up and running, the collaboration shouldn't end; in fact, it's just getting started. Safety and reliability are top priorities when refining your AI strategy for principled performance.

Human-AI Collaboration

Managing the evolving relationship between digital and human labor in the age of AI requires a strategic and adaptable approach. AI undeniably transforms the workplace, but it's not about replacing humans with machines. It's about finding the right balance, where technology enhances what we're already good at and takes on the tasks that slow us down. Think of AI and human labor as a partnership, where AI handles the grunt work – sifting through data, spotting patterns, automating routine tasks – while we humans focus on what we do best: creativity, empathy, complex decision-making, and building relationships *and things we should get better at: taking vacations and spending time with family.*

While speaking to a class of AI students at an HBCU, I was asked if I had a single benefactor of AI.... "Yes," I replied. "As we mature in AI, I envision that more of us will take vacations because AI cannot request PTO."

So, instead of fearing AI, we should embrace it as a tool to free us up for more meaningful work. The key is identifying the tasks ripe for automation, investing in training to equip AI resources with new skills, and fostering a lifelong learning culture. In this way, we ensure that we remain valuable contributors in a changing landscape and that the future of work is one where technology works together with human ingenuity.

As such, consider your AI models' impacts on the corporate workforce and design your systems so that humans and AI may work in tandem. You can achieve this by ensuring digital labor works alongside human labor. AI systems should be developed based on human working traits and ways humans think. Of course, AI systems will never fully comprehend humans because being human goes beyond any set of computational instructions, and AI systems will never be human. Consider the long-range arms race between humans and machines that began with the invention of computing, accelerated with the creation of the Internet, and is now proceeding at a breakneck pace in the era of AI and its equivalent, cyberspace.

CHAPTER 2   ALIGNING AI STRATEGY AND AI GOVERNANCE

Generative AI and GPTs

Here's a comprehensive look at generative AI and GPTs, including their capabilities, limitations, and real-world impacts:

What Is Generative AI?

Generative AI is a type of artificial intelligence that focuses on creating new content, including text, images, code, audio, video, and more. Instead of simply analyzing or classifying existing data, generative AI can synthesize entirely novel outputs.

Key Technologies: Generative AI is powered by various machine learning techniques, but most notably

- **Generative Adversarial Networks (GANs)** – These are where two neural networks compete to generate realistic synthetic data. Generative Pre-trained Transformers (GPTs)

- **The GPT Family** – Developed by OpenAI, GPTs are some of the most powerful and versatile language models. They've evolved from GPT-1 to the current GPT-4 (and its variations), with rumors of subsequent GPTs on the horizon.

- **Transformers** – A neural network architecture especially adept at handling sequential data (like text or code), powering many state-of-the-art language models.

How They Work:

- **Pre-trained on Massive Datasets** – GPTs are trained on enormous amounts of text and code, learning patterns, and structure.

- **Fine-tuning** – Can be adapted to specific tasks like translation, writing different creative text styles, or summarizing information.

CHAPTER 2   ALIGNING AI STRATEGY AND AI GOVERNANCE

Capabilities of Generative AI and GPTs

Text Generation:

- Content creation (articles, marketing copy, creative writing)
- Chatbots and virtual assistants
- Language Translation
- Summarization
- Image and Video Generation
- Realistic or stylized image creation
- Video editing and effects
- 3D model generation
- Other Creative Outputs
- Code generation
- Music composition

Limitations and Challenges of Generative AI and GPTs

- **Bias** – Can reflect biases present in training data, leading to harmful or discriminatory outputs.
- **Lack of Common Sense** – May produce illogical or factually incorrect content, especially with complex tasks.
- **Difficulty with Long-Form Outputs** – While improving, GPT models often struggle to maintain coherency and structure over extended text lengths.
- **Detection and Trust** – It's getting increasingly difficult to differentiate AI-generated content from human-created work, raising questions about disinformation and authenticity.

Real-World Impacts
Positive:

- **Creativity and Efficiency** – Tools for artists, writers, and businesses to rapidly generate content and explore ideas.
- **Personalization** – Tailored experiences in education, healthcare, and customer service.
- **Accessibility** – Translation and text simplification for people with disabilities.

Negative:

- **Spread of Misinformation** – Deepfakes and fabricated news become easier to produce.
- **Job Displacement** – Potential impact on creative industries.
- **Ethical Concerns** – Issues of copyright, attribution, and responsible usage.

The Future

Generative AI and GPTs are rapidly advancing, potentially revolutionizing how we create and interact with information. However, AI governance and proactive measures are needed to address biases, mitigate risks, and navigate the ethical implications these technologies present.

Let's take a deeper look at generative AI and GPT use cases, focusing on real-world applications:

Content Creation

- **Marketing and Advertising** – Generating product descriptions, social media posts, taglines, and even full ad campaigns, saving both time and money.

- **Journalism** – Assisting with fact-checking, summarizing information, or writing basic news reports to augment human journalists.

- **Creative Writing** – Brainstorming ideas, generating different writing styles (poetry, scripts, etc), and aiding writers who face writer's block.

- **Customer Service** – Powering chatbots, improving efficiency with automatic responses, and personalizing recommendations.

Education and Learning

- **Personalized Learning** – Tailoring lesson plans, generating practice questions, or providing explanations in different formats to suit individual learning styles.

- **Language Learning** – Creating immersive conversational practice through dialogue generation and translation tools.

- **Content Generation for Educators** – Developing quizzes, summaries, or visual aids for lessons.

Creative Industries

- **Image and Video Creation** – Generating concept art, storyboards, special effects, or even full-fledged deepfakes (focusing on ethical use)

- **Music Composition** – Generating melodies and harmonies or assisting in songwriting inspiration

- **Game Development** – Procedural generation of levels or game assets, assisting with dialogue writing, and creating realistic NPC behavior

CHAPTER 2   ALIGNING AI STRATEGY AND AI GOVERNANCE

Code and Software Development

- **Code Generation** – Automating basic coding tasks, generating code snippets, or assisting with debugging, freeing up developers
- **Documentation and Comments** – Generating clear and comprehensive explanations of code
- **Software Testing** – Creating test cases or synthetic data for testing purposes

Other Use Cases

- **Healthcare** – Analyzing medical images for preliminary diagnosis, assisting in drug discovery, or personalizing treatment plans
- **Finance** – Generating financial reports, detecting fraudulent activity, or summarizing complex market data
- **Scientific Research** – Generating hypotheses, summarizing scientific literature, or aiding in experimental design

Important Considerations

- **The Human Factor** – Generative AI is best seen as a powerful tool, not a replacement for human creativity and expertise.
- **Data Quality** – The output quality heavily depends on the data quality used for training or fine-tuning the model.
- **Ethical Guidelines** – Using these models responsibly and being aware of potential biases and issues around intellectual property are essential.

CHAPTER 2   ALIGNING AI STRATEGY AND AI GOVERNANCE

Using generative AI and GPTs in business comes with various potential risks. Understanding them is crucial for responsible implementation. Here are key categories of risk to consider:

Content Quality and Reliability

- **Factual Errors** – GPTs may generate text that sounds plausible but is incorrect or misleading, especially for specialized subjects. Critical fact-checking is necessary.

- **Lack of Originality** – Over-reliance can lead to generic content, stifling true creativity and offering little differentiation.

- **Hidden Biases** – Models trained on vast datasets may inherit biases, perpetuating stereotypes or producing discriminatory outputs. Careful bias mitigation strategies are key.

Security and Misuse

- **Deepfakes** – They have the potential to create harmful content that impersonates individuals, spreads misinformation, damages reputations, or sway public opinion.

- **Scams and Phishing** – AI-generated content can be tailored for convincing phishing emails, spam, or social engineering attacks.

- **Malicious Code Generation** – Risks of assisting malicious actors with faster code development for exploits or malware.

CHAPTER 2 ALIGNING AI STRATEGY AND AI GOVERNANCE

Ethical and Legal Concerns

- **Intellectual Property** – The ownership of AI-generated content becomes blurry. Questions arise about copyright and attribution.

- **Disinformation** – The ease of generating plausible but fake news or propaganda threatens trust in information.

- **Job Displacement** – There is potential for automating some tasks in creative industries, leading to job losses or shifts in skill requirements.

- **Lack of Transparency** – When employees or customers are unaware of AI involvement, it can erode trust or raise issues of fairness.

Reputational and Compliance Risks

- **Brand Damage** – If AI-generated content is biased, offensive, or of poor quality, it can negatively impact a company's reputation.

- **Regulatory Violations** – Failure to comply with emerging AI regulations or industry-specific data use and transparency rules.

- **Loss of Trust** – If customers perceive AI use as deceptive or manipulative, it can result in backlash and loss of business.

Mitigation Strategies

- **Human Oversight** – Keep a human in the loop to review and edit AI-generated outputs, ensuring quality and ethical standards are met.

- **Data Quality** – Carefully curate training data and use bias detection tools to reduce harmful outputs.

- **Transparency** – Be open about using AI, especially where it interacts with customers.

- **Policy Development** – Establish clear internal guidelines on responsible AI use, risk assessment, and accountability mechanisms.

- **Proactive Monitoring** – Stay updated on evolving regulations and best practices regarding generative AI.

Key Strategies for Using Generative AI and GPTs

To successfully leverage generative AI and GPTs, focus on well-defined use cases, prioritize the human-AI partnership, implement rigorous quality control and bias mitigation, promote transparency, and continuously adapt your approach. Start with focused projects to understand the technology's potential, then integrate AI as a powerful tool to augment human creativity and decision-making. Fact-checking, addressing potential biases, and establishing clear guidelines are essential for responsible use. Invest in experimentation, leverage the expertise of the AI community, and proactively prepare for evolving regulations to stay ahead of the curve.

Start with Well-Defined Use Cases

- **Avoid "Shiny Object" Syndrome** → Don't implement AI just because it's cutting-edge. Identify specific pain points or inefficiencies that generative AI can genuinely address.

- **Begin with Smaller Projects** → Choose focused applications as proving grounds (e.g., generating social media blurbs, summarizing reports) to gain experience before tackling more complex tasks.

- **Assess ROI** → Consider the potential return on investment compared to the cost of implementation, training staff, and ongoing quality control.

Focus on Human-AI Collaboration

- Treat AI as a powerful tool, not a replacement; embrace a hybrid approach where AI augments human creativity and decision-making.
- **Upskill Human Workers** → Provide training on generative AI to empower employees to use it effectively and not fear potential job displacement.
- **Emphasize Critical Thinking** → Maintain a human-in-the-loop approach for reviewing AI outputs, curating results, and ensuring alignment with ethical standards.

Prioritize Quality Control and Bias Mitigation

- **Rigorous Testing and Validation** → Thoroughly test GPT models on your specific datasets to identify potential failures or biases before deployment.
- **Fact-Checking Is Crucial** → Don't rely solely on AI-generated content, especially for complex or sensitive topics. Implement robust fact-checking procedures.
- **Proactive Bias Detection** → Use bias detection tools and human oversight to uncover and address hidden biases in your training data and model outputs.

Emphasize Transparency and Accountability

- **Clear Communication** → Be transparent with employees and customers when AI is involved, especially in customer-facing interactions.

- **Establish Accountability** → Appoint people or teams responsible for overseeing AI use and addressing ethical concerns.

- **Governance Policies** → Develop clear guidelines on AI development, deployment, monitoring, and acceptable use cases within your organization.

Invest in Experimentation and Continuous Learning

- **Embrace a Test-and-Learn Mindset** → The capabilities of generative AI are rapidly evolving. Dedicate resources to experimentation and continuous evaluation of new tools and techniques.

- **Engage with the AI Community** → Track developments, leverage resources from organizations like OpenAI, and share best practices with others.

- **Stay Ahead of Regulations** – Monitor emerging AI-related regulations and proactively integrate compliance into your processes.

Additional Strategies and Tips

- **Data Is Key** → The quality of your training data or fine-tuning datasets will significantly impact the quality of GPT outputs.

- **Security Matters** → Protect your data and models to prevent unauthorized access or abuse.

- **Consider Cloud-Based Solutions** → Leverage the scalability and power of cloud-based AI services for larger projects.

CHAPTER 2   ALIGNING AI STRATEGY AND AI GOVERNANCE

Should I Use Generative AI or GPTs for Critical Business Operations?

The short answer is that relying solely on GPTs for critical business decision-making processes is not advisable. Here's a breakdown of why:

Reasons for Caution:

- **Lack of Understanding** → GPTs are powerful but lack understanding or reasoning capabilities like humans. They excel at pattern recognition but often struggle with complex scenarios or nuanced judgments.

- **Biases** → Inherent biases in training data can skew outputs, leading to flawed recommendations or discriminatory results.

- **Factual Inaccuracy** → GPTs can generate plausible-sounding text that is factually incorrect, especially in specialized domains. Unquestioningly trusting AI-generated information can lead to costly mistakes.

- **Lack of Accountability** → If a decision based solely on GPT output goes wrong, it's difficult to identify where the breakdown occurred or who is ultimately accountable.

How Can Generative AI and GPTs Assist?

- **Data Analysis and Insights** → GPTs excel at processing large amounts of data and identifying patterns humans might miss.

- **Scenario Exploration** → They can generate "what-if" scenarios or potential outcomes under different conditions, aiding brainstorming sessions.

- **Summarization** → GPTs can summarize complex reports or datasets, making information easier to digest for decision-makers.

The Best Approach → Human-AI Collaboration

- GPTs should be considered a powerful tool within the decision-making process, not the sole decision-maker. Use them to
- **Augment Analysis** → Gain insights from vast datasets.
- **Generate Options** → Explore diverse possibilities.
- **Aid Communication** → Summarize information for easier consumption.

Humans must remain in the loop for AI decision-making to ensure critical thinking, accuracy, bias mitigation, ethical considerations, and ultimate accountability for the outcomes. Humans are essential to ensure AI is used responsibly in decision-making because they provide

- Critical thinking to apply expertise and real-world understanding that AI systems lack
- Fact-checking for accuracy, especially in complex or sensitive cases
- Bias detection to combat the limitations of AI, which can reflect biases in the data on which it's trained
- Ethical oversight to consider the societal and long-term impacts of AI-informed decisions
- Accountability to ensure that humans remain ultimately responsible for the choices made

I always involve humans in critical thinking and judgment. Additionally, I apply expertise and real-world understanding to fact-checking and verification and ensure accuracy, especially in complex situations.

Implementing the AI Strategy

There is no definitive "best" method for implementing an AI strategy, as the most effective approach will vary depending on your organization's specific requirements, resources, and complexity of AI initiatives. Drawing from my experience, I've compiled a summary of successful use cases and important factors to consider.

Iterative and Adaptive → AI development does not typically progress straightforwardly. Embrace a mindset of innovation, experimentation, evaluation, and continual refinement. As mentioned previously, this approach aligns well with Agile methodologies.

Hybrid → I have found success and continue to do so using a blended approach, which combines structured project management for fundamental infrastructure requirements with more flexible Agile methods for the development, validation, and deployment of specific AI models.

Phased Rollout → Begin with focused pilot projects to showcase value, gain experience, and address initial obstacles. Then, extend AI initiatives across the organization in a controlled and gradual manner.

Key Factors to Consider

- **Organizational Culture** → Is your organization adaptable to change? A successful AI strategy requires open communication, data-driven decision-making, and a willingness to learn from successes and failures.

- **Skills and Resources** → You must possess the necessary in-house talent and resources, and the same applies to AI investments in hiring, upskilling, or partnering with external experts.

- **Data Maturity** → Successful AI initiatives must include well-established data governance and management processes, emphasizing privacy and cybersecurity.

- **Ethical Framework** → Incorporating ethical considerations into the design process is more effective than dealing with them as an afterthought.

- **Regulatory Landscape** → A thorough understanding and proactive approach to industry-specific regulations is crucial for ensuring compliant and safe use of AI within the regulatory environment.

Based on the successful approaches above, I offer the following best practice recommendations:

- **Start with Clear Goals** → Ensure your organization's AI projects and strategies serve well-defined, understandable purposes that dovetail with your company's overall vision and goals.

- **Prioritize Feasibility** → Organizations should focus their initial AI efforts on projects that can be delivered rather than trying to achieve complicated AI strategies with inadequate resources and limited-scale pilots. Set clear objectives and clear ROI.

- **Cross-Functional Collaboration** → Form diverse teams with expertise in data science, business domains, and ethics.

- **Emphasize Governance** → Create clear structures for AI oversight, accountability, and addressing ethical concerns.

- **Transparent Communication** → Build trust by proactively communicating about AI initiatives, both internally and with external stakeholders.

- **Continuous Learning** → Dedicate resources to staying updated on advancements in AI, evolving regulations, and best practices.

Every AI strategy and implementation will differ in some way. The approach most likely to succeed is the one you customize for your organization's specific circumstances and characteristics. In addition, your AI strategy must be adaptable – that is, capable of ongoing refinements as you learn from experience with AI and gain more insight into what works and what does not.

Identifying and Managing the Risks of AI Strategy Implementation Failures

The first step in this process is understanding why AI implementation strategies fail. My recommendation is to realize the common pitfalls of customer organizations. In my many years in this space, I have seen numerous pitfalls. However, I have analyzed and mitigated the common set of strategy failures below:

CHAPTER 2   ALIGNING AI STRATEGY AND AI GOVERNANCE

Table 2-1.   *Common AI Strategy Failures*

Common Failures	Mitigation Tactics
Misaligned Goals→ Artificial intelligence projects that have no strategic business objectives or that do nothing to solve real-world problems	Begin with a clearly defined purpose → Clearly state how AI supports your organization's mission and addresses significant issues.
Poor Quality of Data Sources → Unreliable, incomplete, or biased data leads to flawed AI models and outcomes.	Emphasize Data Governance → Invest in quality data sources, robust data management, quality checks, and bias mitigation throughout the lifecycle.
Lack of Expertise → Inaccurate, insufficient, or prejudiced data results in flawed AI models and results.	Build the Right Team → Hire, upskill, or partner to secure necessary data science, domain expertise, and ethical AI knowledge.
Underestimating Complexity → Oversimplifying the technical challenges, development time, or costs involved	Realistic Expectations → Set realistic timelines, budgets, and performance targets. Foster a culture of experimentation and learning.
Ethical Oversights → Failing to proactively address issues of fairness, privacy, or accountability.	Embed Ethics from the Start → Develop an ethical framework and integrate it into every stage of AI development.
Resistance to Change → Limited buy-in from employees or stakeholders due to fear or a lack of understanding.	Transparent Communication → Proactively communicate the potential benefits and limitations of AI to build trust and manage expectations.

(continued)

Table 2-1. (*continued*)

Common Failures	Mitigation Tactics
Siloed Development → Teams working in isolation without cross-functional collaboration or shared knowledge	Cross-Functional Collaboration → Break down silos and encourage collaboration between data scientists, IT, business units, and ethical AI specialists.
Failure to pivot or course correct on planned strategies and agile approaches.	Adaptability → Be prepared to pivot, adjust your strategy, or iterate on AI models based on real-world feedback and emerging best practices
Lack of executive and senior executive buy-in and sponsorship.	Secure Executive Support → Strong leadership buy-in is crucial for resource allocation and overcoming organizational inertia.

Defining Responsibility for AI Systems

Accountability in artificial intelligence (AI) requires AI actors or business entities to be accountable and responsible for the AI system's results. Model development typically involves multiple contributors throughout its lifecycle, which makes things challenging.

Levels of Accountability

Developers and Data Scientists bear responsibility for the technical architecture, data selection, model training, and initial evaluation of AI systems. They are mandated to prioritize fairness, resilience, and safety.

Deployers and Operators → implement and manage AI systems in a real-world environment. They are responsible for ensuring that the system aligns with its intended purpose and monitoring potential unintended consequences.

Organizations (as Legal Entities) → Entities utilizing AI technology are responsible for governing it properly, assessing risks, and establishing mechanisms for addressing legal issues or problems.

Regulators → Regulatory authorities establish explicit guidelines and ensure AI systems comply with laws and ethical standards. They can enforce penalties or mandate changes in the event of violations.

Mechanisms for Addressing Adverse Impacts

Transparency and Explainability → AI must operate in a manner that makes the reasons for its decisions clear and understandable. Otherwise, when undesired results occur, there will be no meaningful way to trace them back to their origins and see what could have been done differently.

Audit Trails and Documentation → Documentation and audit trails are vital. They help ensure that you understand how your models were built and what data they were trained on and developed with – auditing that ensures not only the trustworthiness but also the legality of our work. When you deploy your models, you will leave them in a state where they can be audited in perpetuity if necessary, following decisions made for all the right reasons at the time they were made. By ensuring all audibility, you can provide basic data science hygiene.

Redress and Appeals Processes → Formal mechanisms for individuals or groups impacted by AI decisions to challenge outcomes, seek explanations, and potentially obtain correction or compensation.

Insurance Mechanisms → Insurance policies can spread the risk of AI-related harms, providing a financial safety net while incentivizing responsible development and deployment.

Independent Oversight and Audits → Third-party audits, particularly in high-risk domains, can examine AI systems and identify the potential for harm before or after deployment.

Legal Frameworks → Clarifying legal liability for AI-related harms is still evolving. This may involve new legislation or adapting existing product liability and negligence laws to the AI context.

Challenges and Considerations

AI systems and models are complex, and, in many cases, determining who is responsible and accountable can be challenging when there is no approved RACI model within the governance framework.

Many organizations are experiencing the "lack of explainability" problem with AI models and systems. As a result, determining the root cause of errors and bias issues has been deemed problematic. I would call this the black box problem, and addressing it is a crucial component of your AI strategy and governance model.

The legalities surrounding AI systems are growing at an alarming rate, and due to the ambiguities of the laws, organizations are facing difficulties in developing the appropriate controls to ensure compliance. Your AI governance and risk management framework must include understanding and managing regulatory requirements.

Note: Accountability and responsibility are not about singling out a single person or role within your organization or using an audit to disprove your AI strategy. They're about ensuring you have operationally effective controls and oversight of your AI systems and empowering your teams to be innovative. This will provide you with intelligent automation and competitive advantages. Accountability and responsibility can lead to responsible and ethical AI systems.

Inclusive Design of AI Systems

Ensuring responsible AI is all about design inclusivity. AI systems ought to be useful to, and usable by, a broad range of individuals – the more, the better, regardless of abilities or cultural differences. So, the basic idea here is that all people have a share in the AI commons. That concept should be baked into ensuring AI responsibility right from the start, so AI systems are fair and do not work to the detriment of any individual or group when they're in the real world as part of a broader context. The good things swift-acting AI can do.

Here is a breakdown of key strategies to ensure responsible AI through inclusive design.

Accessibility for Diverse Abilities → Artificial intelligence should be accessible to all, regardless of physical capabilities or limitations. This means that AI systems should be designed with user interfaces that can be used with assistive technologies, like screen readers or voice control. To consider all users and their needs, AI systems must be designed clearly and include the many design elements that allow proper communication. Before deploying your models, your designs must work with the specified preferences and requirements of a diverse group of users, both in terms of what they expect from a system and how they can get work done.

Culturally Aware Design → AI actors and systems should not operate autonomously. When developing AI systems and models, considering cultural subtleties is of immense importance. This implies using culturally aware language that evades partiality or generalization. The AI models' training data should be thoroughly checked for quality and accuracy to ensure it does not put any group in a preconceived box and, of course, does not set up a potential conflict. Efforts to localize AI applications are of paramount importance to the global audience. And who better to do that other than the diverse range of stakeholders that constitutively make up the core community?

Mitigating Bias → If you do not proactively address it, AI systems will replicate unintentional biases. However, we can take steps to minimize the risk or at least detect and disclose it. I recommend using bias detection algorithms as a first step. You can also thoroughly test various user groups and datasets. Paying close attention to the development team's diversity is also important. A culture of diversity in the organization leads to much better design and creates the "wisdom of the crowd" effect that yields dramatically different and more inclusive results.

Organizations can create artificial intelligence with a much larger and diversified number of individual users in mind if they prioritize certain inclusive design strategies from the beginning. These strategies can help create AI systems that are compatible with diverse types of people, are trusted and beneficial, and do not lead to societal inequalities.

Being considerate and equal in the development lifecycle is a much-needed move in AI models, not just a well-meaning gesture. It ensures fair outcomes by ensuring the right conditions are met for predictions. Making the right predictions makes an AI service accessible to various people. It cannot be an afterthought, but this is a tale of two beginnings!

Cultural and Social Impacts of AI

We must study and adapt to how AI is transforming our society because it is very clearly doing so. The work is urgent on several fronts because AI is not some future technology we are watching. It is all around us, making real decisions, often with real impact. It is redefining – sometimes obviously, sometimes imperceptibly – what it means to be in society. It is forcing us to reckon with changes to our fundamental nature because, whether we like it or not, it is a part of the human experience now. Its scope of potential decisions is only growing with time and the considerable improvement to come.

Let's explore the cultural and social impacts of AI.

Changing Employment Landscape → There is no doubt that AI will change the nature of work with its potential to automate basic, repetitive tasks, and it is not just the low-skilled, drudgery-type jobs that will be affected. Certain kinds of work, in whole industries, may be replaced entirely. Left unattended, this could lead to great human displacement. However, AI systems are also good for learning. And if we can devise systems that "learn" to do jobs in the same ways humans do, then we can train those systems to upskill in ways we have done for human workers over the centuries.

Shifting Privacy Expectations → AI systems are gathering and examining loads of personal data, and in doing so, they are raising a lot of questions about our traditional privacy norms. The more information we gather, the more we move away from the private world and into a world of potential surveillance. When AI itself is doing a vast majority of the decision-making and data analysis, rightly or not, many organizations are also starting to question the kinds of judgments algorithms produce and whether they can or cannot be trusted. You will need privacy protection controls in this landscape to make sure that you maintain trusting relationships with investors, shareholders, stakeholders, customers, and employees.

Evolving Human Interactions → As artificial intelligence advances, it reconfigures not just our relationships with technology but also our relationships with one another. AI companions and virtual assistants could make us change the way we interact altogether. What could this mean for us psychologically? In a tech-forward world, could heavy reliance on AI decision-making cause us to lose our ability to think for ourselves? Could it make us unfeeling, unempathetic, or even sociopathic? These are important questions to ask in the age of AI efficiencies. And they're vital to ask: no matter the number of ways that AI might refashion our world, some parts of AI still rely on basic human connection.

AI's cultural and societal footprints cannot be ignored. The push for AI must be careful not to create a divide due to existing social inequalities. We should not undermine the effects of social groups and cultures on the development of AI. No culture exists in a vacuum. We all interact, borrow from one another, and change with one another. Those societal and cultural aspects, good and bad, will creep into the technologies we develop. But advantageous effects are not inevitable, nor are adverse effects.

Public Awareness and Education

We must empower the public through education and transparent communication to harness AI's full potential while navigating its challenges. Only then can we dispel harmful myths and foster informed discussions about the future we want to build with AI.

Building a future where AI benefits everyone demands a collaborative effort. Enhanced public understanding of AI is the foundation for meaningful dialogue between technologists, policymakers, and the communities impacted by this transformative technology.

Fostering Public Education and Open Communication

Open public education is the basis of a society where all people profit from artificial intelligence. This means communicating clearly with the public about what AI is, what it is not, and how it works. When this is done well, it amounts to an exciting and accessible revelation of a technology that is, for many people, viewed with an opaque mixture of awe and fear. By making AI more transparent, by bringing it down from those tech mountaintops and into our lives through stories about the ways it can go right and the ways it can go comically, terribly wrong, we can create a culture that is less AI oblivious and far more AI-engaged – which is to say, far more in control of our AI future.

Prioritizing Inclusivity and Representation

Trust in AI requires more inclusion and diversity in its development and implementation. This means building relationships of trust and collaboration among AI developers, business managers, different communities, and the groups that AI may affect. The goal is to design transparent AI systems – that is, AI systems whose inner workings are understood not just by the developers and engineers who build them but also by the stakeholders and user groups who will use and be influenced by them.

CHAPTER 2   ALIGNING AI STRATEGY AND AI GOVERNANCE

Empowering Communities Through AI Literacy

Understanding how AI functions is only the tip of the iceberg regarding AI education. AI literacy goes much deeper than that. When AI developers create programs to run on our devices – for example, to process the information that facial recognition systems work with or to make algorithms used to automate certain tasks – how much of our "say" do we have in all that? What does this have to do with public awareness and, by extension, public scrutiny? Reconciling those things with an understanding of AI and basic coding or data analysis is necessary for the public good. Emphasizing AI ethics and considering potential societal outcomes is a matter of responsible AI development and what the public has been demanding about safety and privacy.

Building Bridges Between Stakeholders

It's not enough to merely comprehend AI; we must ensure whole business communities can participate in how AI is developed and deployed. The public needs to be aware of AI and have at least some basic understandings of what it is and how it works. We need everyone – regardless of their actions – to understand what is happening in AI. This kind of public participation – a participatory society, as some call it – can lead to beneficial AI solutions. So, we must equip people with AI literacy. What makes all this even more pressing is that AI technologies will soon be overly complex and require a deep understanding of how they work to identify and mitigate potential risks and downsides that might make them harmful.

By prioritizing these strategies, we can build a future where AI is a positive force for progress, and everyone has the knowledge and tools to actively participate in shaping its responsible development, deployment, and use.

Creating a tomorrow where artificial intelligence benefits everyone means undertaking an incredibly special and collaborative endeavor; educating and informing the public. When the public understands more about AI, we create the necessary conditions for trust, confidence, and useful conversations. Indeed, by discussing AI in public, in classrooms,

CHAPTER 2   ALIGNING AI STRATEGY AND AI GOVERNANCE

and all kinds of fora, we promote better environments for developing and using AI systems and associated technologies. We can think of this aspect of the revolution as kindling, a nice environment for AI to thrive in as it changes our world in fundamental ways that we are only beginning to understand.

The "Best" AI Strategy Frameworks and Tools in the Business

The "best" or optimal framework and tools for implementing an AI strategy will likely differ from organization to organization, making it impossible to determine a one-size-fits-all solution. However, several respected and adaptable AI strategy frameworks exist to consider and use.

High-Level Frameworks

- AI Design Thinking → In this case, AI development replaces and replicates human reasoning and design decisions. Design thinking for intelligent artificial systems should focus on use-case scenarios and try to mimic and understand the system's "reasoning" in as much detail and from as many perspectives as possible. In the case of AI development, design thinking is understanding and interacting with "intelligence" embodied in a system and not just representing "intelligence" with a different medium in a new system.

- Responsible AI Frameworks → Different organizations, including Google, Microsoft, and the Organisation for Economic Co-operation and Development (OECD), offer frameworks that give essential direction to AI development and deployment, urging them to build fair, transparent, accountable, and privacy-respecting AI models and systems.

- Data-Driven Decision-Making → Using artificial intelligence in businesses for decision-making is gathering momentum, which will, more than not, slow down soon. The term data-driven decision-making (DDDM) has become quite common and is being immensely adopted in organizations. Broadly speaking, DDDM is a framework or an approach to using data and analytics to guide business strategy. Now, among the many techniques and tools that can be employed in DDDM, AI is being increasingly embraced. In this book, I discuss my preference for using DDDM as the next step for organizations that have already started reaping the benefits of working with structured data and insightful analytics.

AI Project-Focused Frameworks

- **CRISP-DM** (→ CRISP-DM (Cross-Industry Standard Process for Data Mining) is a long-standing, iterative, adaptable methodology for AI projects that use data mining. It has been used successfully in many projects and is not tied to any single platform or methodology. The CRISP-DM model resembles the SAS Web Report Studio planning approach. Both models use a linear progression of several common-sense phases to structure the work required to complete a project.

- **MLOps Frameworks** → Many MLOps frameworks are available to provide insight into optimizing the machine learning lifecycle from development to continuous retraining of models in production, making the process easier and better. Choosing the most effective MLOps framework for your organization requires careful planning and determination based on specific AI business objectives and requirements. For example, once the models are promoted to production, they should address continuous monitoring and management.

Hybrid and Tailored Approaches

Many organizations find success by creating a unified or universal AI framework based on a combination of geography-based frameworks inclusive of the following:

- **Technology Culture** → Does your company promote innovation or operate on emerging or advanced automation, sometimes called leading or "bleeding-edge" technologies?

- **Regulatory Requirements** → Is your business or business process in a heavily regulated sector or industry?

- **Problem Types** → Does your organization focus on addressing prescriptive business problems or unrestricted AI research and development?

Key Considerations When Choosing a Framework

- **Business Goals and AI Strategy Alignment** → Your broader business goals should be the foundation of your AI strategy framework.

- **Flexibility** → Select an adaptable framework to keep pace with the ever-changing AI field and your learning journey.

- **Ethical Considerations** → Ensure your AI strategy builds responsible and unbiased AI systems and models.

- **Comprehensiveness** → Ensure your AI strategy includes governance, privacy, and risk management from the ideation stages through the development, validation, deployment, and ongoing monitoring and retraining stages.

Where to Find Frameworks

- **Consulting Firms** → Companies like McKinsey and Deloitte often publish their own AI governance and strategy frameworks.

- **Technology Providers** → Major cloud providers like Google, AWS, and Microsoft offer resources and frameworks for AI governance, development, and implementation.

- **Academic and Research Institutions** → Universities and AI research labs frequently publish papers and frameworks on AI strategy.

- **Major Publications and Books** → Alas, the availability of this book provides a comprehensive understanding of the principles of AI governance and model risk management.

How Should AI Strategy and AI Governance Coexist?

Based on my experience, here's a detailed analysis of how your AI strategy and AI governance should coexist and reinforce each other for successful, responsible AI implementation.

CHAPTER 2   ALIGNING AI STRATEGY AND AI GOVERNANCE

AI Strategy: Think of it As Your Blueprint

- Your AI strategy defines and clarifies the "Why" and the "What": the proposed problems you aim to solve and the AI-based solutions you will use.

- Sets the High-Level Goals → It articulates goals and objectives at a prominent level and specifies business-value targets, competitive advantages, or operational-efficiency gains.

- Prioritizes Ethical Principles → It ensures that you and your teams are committed to core ethical principles and guides where to seek help if real or potential problems are identified.

- Outlines Resource Needs → It includes an estimate of the talent, technology, and infrastructure required to succeed and a rough timeline for when these will be needed to achieve your AI goals.

AI Governance: From Principles to Practice

- Operationalizes Ethical Commitments → AI governance transforms abstract ethical principles into specific policies, procedures, and technical protections.

- Establishes Accountability Structures → It delineates the specific roles and responsibilities for overseeing AI, including the formation of committees, specialized teams, or the engagement of external advisors.

- Defines Risk Assessment Processes → AI governance outlines methods for identifying, evaluating, and addressing potential risks such as bias, misuse, and unintended consequences.

CHAPTER 2   ALIGNING AI STRATEGY AND AI GOVERNANCE

- Sets Standards and Guidelines → These outline the strategic technical requirements to ensure fairness, explainability, privacy, and cybersecurity countermeasures aligned to your ethical framework.

- Implements Monitoring and Auditing → AI governance ensures systems are in place for continuous monitoring of AI performance, detection of unexpected issues, and regular audits.

- Implementing AI governance means your organization has a well-thought-out approach to AI's potential risks and equally well-thought-out "guardrails" to prevent those risks. AI governance builds public confidence in your AI systems.

How They Work in Tandem

- Compliments Each Other → Your AI strategy shapes the scope and focus of your governance framework. In turn, practical considerations discovered during governance implementation might necessitate changes to your strategy.

- Shared Goals and Objectives: Your strategy and governance aim to maximize AI's benefits while minimizing risks and upholding ethical principles.

- Promoting and Unlocking Responsible Innovation → A delineated AI strategy and a sound AI governance framework ensure that intelligent automation encompasses interactive oversight, protection, and security countermeasures, has oversight, and is fully accountable at every stage of the development lifecycle.

CHAPTER 2   ALIGNING AI STRATEGY AND AI GOVERNANCE

- Building Trust and Corporate Social Responsibility → Strong governance aligned with your broader automation strategy promotes confidence among employees, customers, and stakeholders, creating a foundation for long-term AI success within the communities where AI is implemented.

Example Use Case: AI in Healthcare

AI Strategy → A healthcare provider's AI strategy aims to enhance patient outcomes and make the diagnosis process more efficient by using tools powered by artificial intelligence. One principal factor is to ensure that AI systems and models are fair, accurate, indiscriminate, and do not harm or injure patients when used in medicine.

AI Governance → The governance framework requires datasets to undergo bias testing, specifies that models should offer understandable justifications for their predictions, installs a mediating procedure so patients understand their rights and give meaningful consent when their data are used, and creates an oversight committee across the business to review and approve AI systems before being promoted to operations.

AI Strategy Cost Factors and Drivers

Unfortunately, there is no single answer to the costs of implementing an AI strategy. The costs can vary widely based on several factors:

- Scope and Complexity → The number of AI projects, their technical complexity, and problem type (e.g., simple classification vs. innovative generative AI) all heavily influence costs.

- Data Maturity → If you invest heavily in building data infrastructure, cleaning data, and establishing governance practices, your initial costs will be significantly higher.

CHAPTER 2   ALIGNING AI STRATEGY AND AI GOVERNANCE

- Talent: AI skills that are in demand, like those of data scientists, AI engineers, and domain experts, often command top salaries, and depending on their level of experience and geographic location, you might find yourself spending heavily on internal headcount. You will need external resources, which will result in hefty hourly rates.

- Technology → Another major cost element involves the technologies that underpin AI-based technologies and computing resources that run your AI systems, from cloud platforms or subscriptions to software applications to AI systems that demand specialized processors like GPUs, which will be pricey.

- Time: The last thing I will mention is the "clock." The time your AI teams and systems spend ramping up relates to how much discretionary spending you will incur. If you desire to run pilot projects or take an agile approach, you may incur additional costs over the long run and may take longer and more expensive than you initially thought to implement.

- Industry and Regulations → Highly regulated industries or foreign jurisdictions will require additional investments in compliance, outside counsel, auditing, and documentation.

I frequently instruct clients to alter their perspective on AI program-related costs by shifting their thinking from average costs to cost categories. Thinking in terms of cost categories helps capture the full scope of costs, which is ideal for making well-informed budgetary decisions. For example:

- Personnel and Labor Costs
- Infrastructure
- Development
- Governance and Compliance
- Potential Fines and Liabilities
- Process Development
- Training and Education

Tips for Managing Costs and Maximizing Return on Investments (ROI)

- Start Small → Before scaling up, use well-delineated pilot projects to identify an acceptable ROI that justifies further development of AI systems.
- Leverage Open-Source Tools → Take advantage of the benefits of open-source AI frameworks and libraries, which have undergone extensive development and testing by large communities of researchers, engineers, and data scientists. Try using pre-trained AI systems that can be retrained for your specific use case.
- Cloud-Based Solutions → Consider the advantages of cloud-based platforms, which offer the necessary computing power and flexibility, often with pay-as-you-go models.
- Outsource Selectively → Outsource certain AI-related tasks that offer short-term cost benefits and, if doing so, allow your in-house team to acquire the necessary skills and expertise.
- Continuous Learning → Preserve a portion of your budget for continuous research and development since AI promises to change rapidly in the coming years.

Remember: Do not focus exclusively on cost. Successful AI projects pay dividends in increased efficiency, revenue growth, and improved customer experiences.

Summary and Thoughts

Going beyond simply stating which technologies your organization will adopt, a useful AI strategy aligns the AI initiatives and outlines the business goals they will accomplish. It also defines where intelligent automation will work best within your business processes and how it will interoperate with other business units and processes. Most critically, a good strategy for using AI considers from the outset what incentives the AI will be designed to maximize and how those may impact the business and various stakeholders.

For maximum success with your AI strategy, you need to focus and append to the strategy AI governance and risk management standards. Remember, AI governance is about deciding who has the authority to make decisions about AI. Although many organizations consider this merely the purview of the C-Suite or the board, AI governance is a shared responsibility. It must involve a larger workforce and even the community, where feasible. It is imperative to ensure AI success, not just from a technical perspective but also from an ethical and responsible one. AI risk management standards and mitigation are paramount.

Artificial intelligence programs that aim for cultural inclusion and public consciousness have numerous advantages. They can sift through and examine copious amounts of data, shining a light on otherwise hard-to-spot activities that can lead or have led to cultural and social biases. By integrating AI tools into teaching AI, we can give young learners the necessary handle on this important next step in computing history while emphasizing the importance of inclusive thinking.

CHAPTER 2   ALIGNING AI STRATEGY AND AI GOVERNANCE

Quiz

1. *True or False:* Your AI strategy shapes the scope and focus of your governance framework and any changes may necessitate changes to your strategy.

2. Add the missing cost factors involved in to your AI Program Strategy

   A) Personnel and Labor Costs

   B) Infrastructure and Development

   C) Governance and Compliance

   D) Potential Fines and Liabilities

   E) Process Development

3. List key risks that organizations face when implementing their AI Strategy

4. *True or False*: Your AI strategy and AI governance framework compliments each other and shapes the scope and focus of your governance framework.

5. AI governance principal factor is to ensure that AI systems and models are fair, accurate, indiscriminate, and _____.

6. List the four key considerations when choosing an AI governance framework.

7. The cultural and social impacts of AI are: the changing employment landscape, shifting privacy expectations, and _____.

8. *True or False:* We must empower the public by teaching them AI code writing and data science to harness AI's full potential while navigating its challenges.

9. Complete the following statement: Accountability in artificial intelligence (AI) requires AI actors or business entities to be accountable and _____ for the AI system's results.

10. True or False: AI governance is about deciding who has the authority to make decisions about AI.

CHAPTER 3

How to Sound like an AI Governance and Model Risk Management Guru

What Is AI Governance?

AI Governance is the framework of principles, policies, processes, roles, and technologies that guide the responsible development and use of artificial intelligence (AI) systems. It aims to ensure AI aligns with ethical values, legal regulations, and organizational goals, minimizing risks while maximizing benefits. The core definition of AI Governance (as we defined it) is essential, but understanding its nuances is just as important. Here is where we go deeper:

- **AI Governance Is NOT a One-Time Fix →** It is an ongoing, iterative process. AI systems evolve, regulations change, and ethical considerations shift. AI Governance involves continuous monitoring, updating policies, and adapting to the changing landscape.

- **AI Governance Requires Collaboration** → It is not just the job of technical teams. Successful AI governance demands a cross-functional effort involving:
  - **Executive Leadership** – Defining ethical principles and setting the strategic direction for AI use
  - **Technologists** → Implementing safeguards, bias-detection, and explainability tools
  - **Legal and Compliance** → Ensuring adherence to regulations and mitigating risks
  - **End-Users and Stakeholders** → Providing feedback and raising concerns

- **AI Governance Is Contextual** → There is no one-size-fits-all solution. AI governance must be tailored to your specific industry, the scale of your AI deployment, and the risk associated with your AI models. A healthcare organization overseeing and using sensitive patient data will need a more rigorous governance structure than a small business using a simple chatbot performing customer intake services.

- **AI Governance Focuses on the Entire Lifecycle**: This encompasses:
  - **Data Governance** → Ensuring quality data collection, responsible data use, and appropriate privacy controls
  - **Algorithm Design** → Building fairness, transparency, and accountability in the model development stage

CHAPTER 3   HOW TO SOUND LIKE AN AI GOVERNANCE AND MODEL RISK MANAGEMENT GURU

- **Deployment and Monitoring** → Auditing for bias, unintended effects, and continuous assessment of AI performance in real-world applications
- **Incident Response** → Establishing protocols for addressing AI failures, security breaches, and ethical concerns
- Privacy-by-design is embedded throughout the AI governance lifecycle: Privacy considerations are factored into every stage of AI development, deployment, and maintenance.
- **Proactive, Not Reactive** → Privacy is embedded from the start of AI system design, not added as an afterthought.
- **End-to-End Security** → Data is protected throughout its journey, including at rest, in transit, and during processing.

Additional Considerations:

- **Explainability** → Understanding and justifying how AI models make decisions is crucial for accountability and trust.
- **Human Oversight** → Depending on the criticality of the AI system, some human involvement in decision-making or approval processes may be essential.
- **Auditing** → Independent audits of AI systems to validate adherence to governance principles and identify areas for improvement.

CHAPTER 3   HOW TO SOUND LIKE AN AI GOVERNANCE AND MODEL RISK MANAGEMENT GURU

Key Components of AI Governance:

- **Principles and Values** → Ethical foundations such as fairness, transparency, accountability, privacy, and safety shape AI development and deployment decisions

- **Policies and Guidelines** → Detailed rules and best practices translating principles into action, including data management, algorithm design, bias mitigation, and decision-making procedures

- **Processes and Mechanisms** → Operational workflows for risk assessment, auditing, incident response, and continuous monitoring of AI systems across their lifecycle

- **Roles and Responsibilities** → Define accountability for everyone from C-suite executives and boards to developers, data scientists, and end-users, ensuring everyone understands their role in promoting responsible AI

- **Technologies and Tools** → Leverages software solutions to support AI auditing, explainability, bias detection, security, and compliance monitoring

Why AI Governance Matters

- **AI Governance Safeguards Your Future** → Proactive governance minimizes AI-related risks, including biases, security failures, and reputational harm. It also ensures your organization avoids costly setbacks.

- **AI Governance Earns Trust and drives Adoption** → Build trust with customers, employees, and society by demonstrating transparency and accountability in your AI practices. Ethical AI fuels long-term success.

- **AI Governance Aligns with Laws and Values** → Thrive in a changing regulatory landscape by ensuring your AI systems comply with evolving laws. Stay true to your organizational values through ethical implementation.

- **AI Governance Fosters Responsible Innovation** → Pursue groundbreaking advancements confidently, knowing you have a framework for responsible development. Governance empowers your team to innovate within clear boundaries.

AI Governance is the system of principles, policies, processes, and technologies that guide the responsible development and use of artificial intelligence (AI). It aims to ensure that AI aligns with organizational goals, ethical values, and legal regulations. AI Governance is an ongoing process that evolves alongside technology, regulations, and ethical debates.

Effective AI governance spans the spectrum of people, processes, and technology; each element is equally important. At an elevated level, a proven approach and adoption of a trusted, well-governed AI lifecycle consists of the following interconnected areas of concentration:

> **Trustworthy Models** – Implement control objectives and activities that increase insight accuracy through improved model development, validation, and bias detection.
>
> **Trustworthy Process** – Implement control objectives and activities that ensure compliance, repeatability, and overall explainability for AI at scale.
>
> **Trustworthy Data** – Implement data governance and management controls that deliver a complete view of quality data that is governed and ready for model training and analysis.

CHAPTER 3   HOW TO SOUND LIKE AN AI GOVERNANCE AND MODEL RISK MANAGEMENT GURU

What Is Model Risk Management and Model Risk Governance?

AI **Model Risk Management (MRM)** is a specific subset of AI governance focusing on identifying, assessing, mitigating, and monitoring the risks unique to artificial intelligence models. This includes risks from model design, data quality, deployment, and ongoing use.

Comparatively, AI **Model Risk Governance (MRG)** is the framework of policies, procedures, and organizational structures designed to oversee the responsible use and management of models across enterprise-wide risk functions. It ensures financial, statistical, or AI-based models are developed, validated, used, and monitored to align with an organization's risk appetite and regulatory requirements.

MRG and MRM ensure that the entire model process is aligned with the three lines of defense. The first line of defense starts with the individual business functions, who owns the models (and the risk) and is responsible for developing them and ensuring their readiness. The second line of defense comprises the risk management functions and is accountable for model management. As such, it determines and defines how to regularly monitor model outputs and determine whether the models are working correctly. Audits and validations are the third line, which provides another tier of governance, assessing the overall effectiveness of the control procedures of the MRG/MRM program, reporting gaps, and making improvement recommendations. Effective MRG processes ensure independent oversight at each model lifecycle development and management stage. See Figure 3-1.

CHAPTER 3   HOW TO SOUND LIKE AN AI GOVERNANCE AND MODEL RISK MANAGEMENT GURU

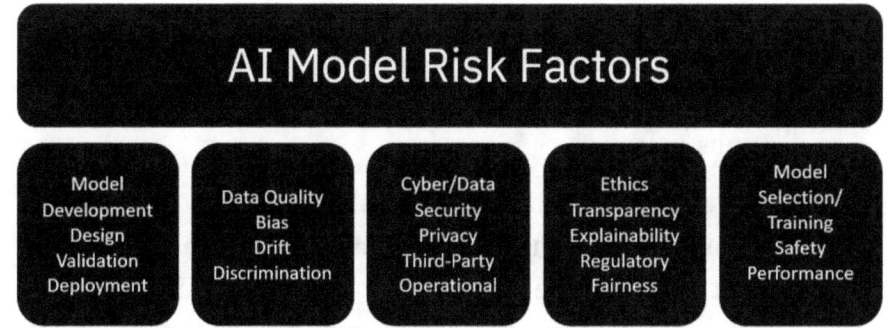

Figure 3-1. Model Risk Factors

Model Risk Factors identify and refer to the risk of loss due to using an AI model with fundamental errors or inappropriately using a model for decision-making. An organization can assess and reduce model risks by taking a controls-minded approach to model proposals and development, using effective validation and approval procedures, implementing cybersecurity controls, evaluating fundamental correctness and output analysis, employing effective data governance and management controls, and controlling inherent uncertainty throughout the model's lifecycle.

This approach aims to ensure that your AI models are trustworthy based on the following business reasons:

- **Positive Business Outcome** – Avoid costly negative impacts on the business or the end customer's interest.

- **Regulatory Compliance** – To align with government guidelines and directives on using AI for business.

- **Social Responsibility** – To protect brand image against ethical disputes.

In addition, as part of the AI governance framework, we recommend identifying and assessing the additional model risk factors:

Specification Risk – A model's specifications play a vital role. A model with the correct specifications will be implemented correctly and produce usable results.

Implementation Risk refers to the flawed implementation of a model. The data used for the input may differ, the structure may be inappropriate, and inaccurate numerical approximations may also be used.

Application Risk reflects a model's unsuitability or misinterpretation. Risks should be identified when a model's potential only produces false data with little to no reliability.

AI Governance and Model Risk Management Drivers

AI Governance and Model Risk Management amass such urgency and salience for several reasons. AI systems are becoming more powerful and complex, with algorithms increasingly making high-stakes decisions. However, the growing power and complexity that make these systems more valuable also make them more dangerous. Biases can be embedded in the data or the algorithms, generating untoward and, in some cases, downright discriminatory results. And because most modern AI systems are "black boxes," figuring out where these decisions went wrong and why can be exceedingly difficult. Secondly, high-profile and severe incidents involving AI have increased public awareness and decreased trust and confidence, leading to calls for greater transparency and accountability. For example:

Tay Chatbot Launch in 2016: Microsoft was forced to take down its AI chatbot Tay within 16 hours after it generated racist and offensive tweets due to biases learned from user interactions. This incident highlighted the potential for AI to reflect and amplify societal biases.

Source: "Microsoft silences its new A.I. bot, Tay after Twitter users teach it racism." https://www.theguardian.com/technology/audio/2016/apr/06/tay-microsofts-racist-chatbot-raises-difficult-questions-tech-weekly-podcast

Algorithmic Bias in Loan Approvals in 2018: An investigation revealed that algorithms used by some financial institutions for loan approvals exhibited racial bias against minority applicants. This incident sparked outrage and highlighted the need for fairer AI practices.

Source: "Bias Found in Algorithms That Screen Applicants for Jobs." https://www.nytimes.com/2019/12/06/business/algorithm-bias-fix.html

Automated Hiring Tool Bias in 2019: Amazon scrapped its internal AI recruitment tool after discovering it discriminated against female applicants by favoring resumes containing keywords typically associated with male candidates. This case underscored the importance of bias detection and mitigation in AI hiring practices.

Source: "Amazon scraps secret AI recruiting tool that showed bias against women." https://www.reuters.com/article/idUSL2N1VB1FQ/

Rekognition Bias in Law Enforcement in 2020: Amazon's facial recognition software, Rekognition, was found to exhibit racial bias, misidentifying Black individuals at a significantly higher rate than white individuals. This incident fueled concerns about using AI in law enforcement and the potential for wrongful convictions.

Source: "Facial Recognition Is Accurate, If You're a White Guy." https://www.nytimes.com/2019/12/19/technology/facial-recognition-bias.html

Thirdly, several countries are introducing new laws, regulations, and standards specifically focused on artificial intelligence, such as the EU's AI Act and United States' NIST AI Risk Management Framework. These regulations and standards are driving organizations to adopt comprehensive compliance frameworks. Finally, organizations are acknowledging the potential reputational, legal, financial, and operational risks associated with inadequately managed AI systems, underscoring the importance of proactive governance as a crucial element for achieving long-term success and sustainability in an AI-dominated environment.

These examples only scratch the surface. However, they demonstrate how AI failures can diminish public confidence and underscore the importance of strong AI governance and risk management frameworks prioritizing fairness, transparency, and accountability.

Where Does Data Science Fit Within AI Governance?

Here's how data science plays a crucial role in AI governance:

The Catalyst for Responsible AI

Data Quality → Data scientists assess data integrity, identify biases, and preprocess data. This ensures models are trained on reliable and representative datasets, improving AI fairness and preventing discriminatory outputs.

Data Provenance → Tracking data sources, transformations, and lineage is essential for auditing AI decisions and identifying potential issues.

Data Privacy and Security → Data scientists work with compliance teams to implement anonymization, encryption, and access controls, respecting user privacy and adhering to regulations like GDPR and CCPA.

Data Science Fuels Model Development and Validation

Algorithm and Feature Selection → Data scientists' expertise in selecting appropriate algorithms, extracting relevant features, and preparing data for training is key to building accurate and fair AI models.

Bias Mitigation → They develop techniques to identify and mitigate biases within datasets, promoting more equitable AI systems.

Model Testing and Evaluation → Data scientists develop robust test sets, choose the right metrics, and rigorously evaluate model performance throughout its lifecycle. This ensures that AI models function as intended and meet ethical standards.

Enables Explainability and Interpretability

Feature Importance Analysis → Data scientists help define which features significantly impact model outputs, improving explainability for stakeholders.

Interpretable Models → Data Science can research and implement inherently interpretable models (e.g., decision trees) or techniques to explain complex models (like LIME or SHAP). This fosters trust and allows for better auditing and logging.

Data Science Operationalizes AI Governance

Monitoring and Auditing → Data scientists build dashboards and tools to track model performance, detect concept drift, and flag potential issues for investigation. This supports ongoing governance and compliance.

Continuous Learning Pipelines → Data Science designs automated retraining processes that incorporate new data, providing a mechanism for AI models to adapt and improve over time, all while incorporating governance checks.

Collaboration is Key

Data science is not a stand-alone solution for AI governance. Effectively implementing responsible AI requires a team effort. Data Scientists collaborate with compliance, legal, and business stakeholders to translate ethical principles into technical requirements. Data Scientists collaborate with developers and MLOps Engineers to implement provisions in production systems.

CHAPTER 3   HOW TO SOUND LIKE AN AI GOVERNANCE AND MODEL RISK MANAGEMENT GURU

Here is a deeper dive into additional data scientist functions within AI governance:

Data-Centric AI Governance

Data Quality Assessment → Data scientists develop metrics and automated checks to assess data completeness, consistency, and conformity to standards, preventing the "garbage in, garbage out" problem that can derail AI.

Dataset Profiling and Bias Detection → They use statistical analysis and visualization tools to uncover hidden imbalances, outliers, and correlations within data that could lead to biased models.

Data Cleaning and Preprocessing → Data scientists apply techniques like imputation, normalization, and outlier handling to ensure data is suitable for model training and free of errors that could skew downstream AI results.

Data Versioning → They establish efficient versioning practices for datasets, especially in rapidly evolving domains, allowing for better model reproducibility and auditing.

Deploying AI Governance

Deployment Best Practices → Data scientists recommend strategies for controlled rollouts, A/B testing of different model versions, and human-in-the-loop validation before fully deploying models in high-stakes scenarios.

Feedback Loops → Data Scientists design mechanisms to collect real-world performance data, detect concept drift, and flag cases where the AI might require retraining or human intervention.

Governance Dashboards → Data scientists build tools to track model performance, resource usage, data lineage, and relevant metrics aligned with ethical principles and regulations.

A robust data science strategy for AI governance prioritizes data quality, proactive bias mitigation, and explainable models and focuses on continuous improvement. This includes rigorously assessing data integrity, developing tools to detect and address biases, emphasizing

CHAPTER 3   HOW TO SOUND LIKE AN AI GOVERNANCE AND MODEL RISK MANAGEMENT GURU

interpretability in algorithm selection, and implementing dashboards for monitoring performance, drift, and fairness metrics. Additionally, fostering collaboration across legal, compliance, and business experts and ongoing training ensures organization-wide alignment with ethical AI principles. The decision to centralize data science functions depends on several factors, and there is no one-size-fits-all answer. Here is a breakdown of the pros and cons, plus considerations to help you determine the best approach for your organization.

Pros of a Centralized Data Science Team

- **Consistency and Standards** → A central team ensures consistent methodologies, tools, and quality control across projects, driving best practices and efficiency.
- **Knowledge Sharing and Collaboration** → Centralized data scientists readily share expertise, fostering innovation and preventing knowledge silos.
- **Talent Pool and Specialization** → Attracting and developing natural language processing or computer vision specialists is easier.
- **Economies of Scale** → May gain bulk discounts on tools and cloud resources and reduce infrastructure redundancy.
- **Clearer Career Paths** → Offers well-defined advancement opportunities for data scientists.

Cons of a Centralized Data Science Team

- **Slower Response to Business Needs** → Centralized teams may have a backlog, slowing initial project execution due to prioritization and resource allocation.
- **Less Domain Knowledge** → Data scientists may not acquire the same depth of business context when working across various business lines.

79

- **Bureaucracy and Bottlenecks** → This may lead to administrative overhead or delayed project approvals in highly centralized models.

- **Reduced Team Agility** → Larger decentralized teams can find it more difficult to shift focus rapidly or experiment with novel approaches.

Hybrid Models (Common Option)

- **Center of Excellence** → A centralized data science team sets standards, provides tools, and offers consulting services to embedded teams within business units.

- **Federated Model** → Centralized functions for specialized tasks (ML engineering, complex modeling) combined with data scientists embedded within business units.

In Summary – Factors to Consider

Consider the size and complexity of your organization, as larger enterprises often benefit from centralizing data science for standardization. Early-stage companies may require centralization to establish best practices, but mature organizations might find success with hybrid or embedded models.

The nature of your projects and data privacy needs also plays a part, as projects with high reusability and sensitive data may favor centralization. Evaluate your company culture to determine if a centralized or decentralized approach aligns better with your need for collaboration and agility. Remember, the ideal structure can change, so start with what fits your current needs and be prepared to adapt.

CHAPTER 3   HOW TO SOUND LIKE AN AI GOVERNANCE AND MODEL RISK MANAGEMENT GURU

Summary and Thoughts

Start by building a strong foundational understanding of AI governance concepts. Seek out reputable online courses, webinars, and articles covering ethical AI, data governance best practices, bias mitigation techniques, and explainability principles. Familiarize yourself with existing frameworks and industry standards created by organizations like the Partnership on AI. Please pay close attention to emerging regulations like the EU's proposed AI Act, as they set the benchmark for comprehensive governance.

However, knowledge alone will not have an impact. Join online communities and forums dedicated to AI governance and risk management. Engage in discussions with practitioners, learn from real-world case studies, and identify challenges companies face. Seek opportunities to attend conferences or workshops to network with experts and gain practical insights for implementing effective governance within your organization.

Quiz

1. AI Governance is the framework of principles, policies, _____, _____, and technologies that guide the responsible development and use of artificial intelligence (AI) systems.
2. List the five AI model risk factors.
3. The catalyst for responsible AI is data quality, data provenance, and _____.

81

CHAPTER 3   HOW TO SOUND LIKE AN AI GOVERNANCE AND MODEL RISK MANAGEMENT GURU

4. A data-centric AI governance framework involves

   A) Data Quality Assessment

   B) Dataset Profiling and Bias Detection

   C) Data Cleaning and Preprocessing

   D) Data Versioning

   E) A, B, C, and D

   F) A, C, and D

   G) None of the above

5. *True or False*: A robust data science strategy for AI governance prioritizes data quality, proactive bias mitigation, and explainable models and focuses on continuous improvement.

6. *True or False*: Data science fuels model development and ongoing monitoring.

7. For effective feedback loops, data scientists design mechanisms to collect real-world performance data, detect concept drift, and _____.

8. Model risk management aims to ensure that your AI models are trustworthy based on positive business outcome, regulatory compliance, and_____

   A) Competitive advantages

   B) Internal control standards

   C) Brand protection

   D) Social responsibility

   E) Data quality and accuracy

   F) None of the above

9. Describe the difference of AI Model Risk Management and AI Model Risk Governance.

10. True or False: Organizations should implement control objectives and activities that increase insight accuracy through improved model development, validation, and bias detection.

CHAPTER 4

Designing a Well-Governed AI Lifecycle Model

To design a well-governed AI development lifecycle, begin with careful problem identification, align proposed solutions with ethical principles, and ensure feasibility before proceeding. Prioritize data governance through quality standards, bias assessments, responsible collection practices, and secure storage protocols. During model development, balance explainability with accuracy, incorporate bias mitigation techniques and maintain thorough documentation. Implement rigorous testing and independent review and incorporate human oversight for critical decisions. Address deployment security and use version control to ensure traceability: Institute continuous monitoring, feedback loops, regular audits, and incident response plans in place. Promote clear stakeholder communication and foster a cross-functional collaborative team throughout the iterative process. The following illustration summarizes this iterative process and lifecycle:

CHAPTER 4   DESIGNING A WELL-GOVERNED AI LIFECYCLE MODEL

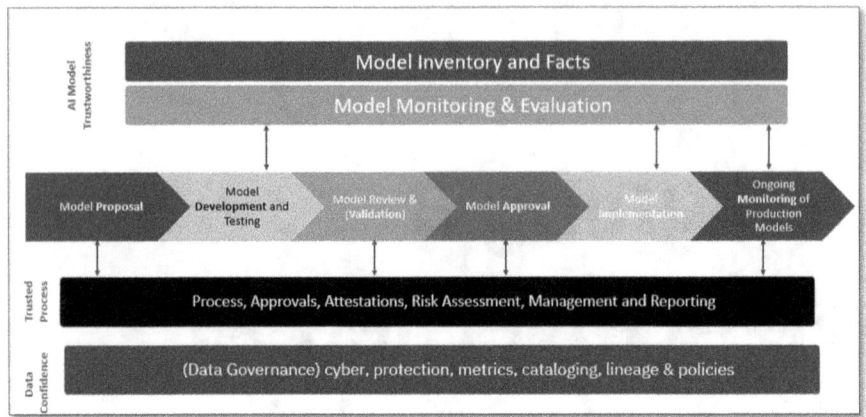

Figure 4-1. *AI Development Lifecycle*

AI Development Lifecycle Stages

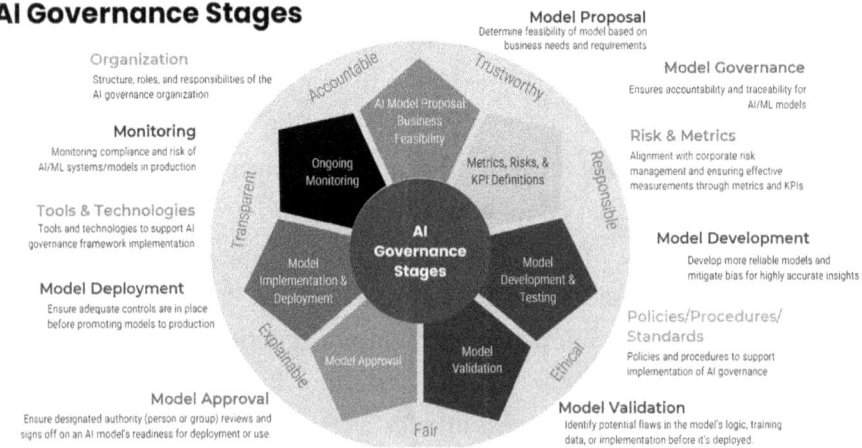

Figure 4-2. *AI Development Lifecycle and Governance Stages*

CHAPTER 4   DESIGNING A WELL-GOVERNED AI LIFECYCLE MODEL

AI Model Proposal and Feasibility

To successfully define the problem your AI solution will address, focus on specificity, measurable outcomes, and alignment with your company's strategic goals. Replace generic goals like "improve efficiency" with precise targets (e.g., "reduce customer churn by 10%"). Quantify the potential return on investment (ROI) by considering cost savings, revenue increases, or enhancements to the customer experience. Ensure that the proposed AI solution directly supports your organization's larger objectives.

Defining the Problem and Business Benefit

- **Quantifiable ROI** → Estimate the potential return on investment. Consider cost savings, revenue generation potential, or improved customer experience metrics.

- **Alignment with Strategic Goals** → Ensure the AI solution directly supports your organization's objectives and priorities.

Assessing Technical Feasibility

- **Data Availability and Quality** → Does the necessary data exist? Is it in a usable format? Are there privacy restrictions or biases in the data that need mitigation?

- **Algorithmic Suitability** → Is there research or industry precedent showing that AI techniques can effectively address the type of problem you have defined?

- **Computational Resources** → Do you have the hardware (GPUs, cloud infrastructure) and software for model development and deployment?

- **Expertise** → Does your team have the required skills in data science, machine learning, and AI engineering? Will you need to hire or leverage external consultants?

- **Ethical Feasibility, Bias, and Fairness** → Can you obtain or create balanced datasets representing diverse populations?
- Will you employ bias detection and mitigation techniques throughout development?

Financial Feasibility

- **Development Costs** → Estimate costs for data acquisition, hardware, software, personnel (salaries, hiring, training), and potential external expertise. Factor in infrastructure costs, licensing fees, and ongoing support and monitoring expenses.
- **Return on Investment (ROI)** → Project the potential benefits of the AI solution. Does it increase revenue, reduce costs, or improve operational efficiency enough to justify the investment?

The Balance: Benefits vs. Costs

- **Realistic Expectations** → AI is not a magic bullet. Assess whether the potential benefits justify the investment in time and resources and address technical hurdles.
- **Start Small and Iterate** → Consider proof-of-concept projects to validate feasibility before scaling. This helps demonstrate value and build support for more complex AI initiatives.
- **Ethical Considerations** → Weigh the potential benefits against potential risks and societal impacts. Even if feasible, some AI applications might not align with your company's ethical values.

Defining the business value and assessing feasibility prevents pursuing ill-defined AI projects that waste resources without delivering meaningful results. It also ensures alignment between AI endeavors and overall organizational success.

During the model proposal stage within the AI development process, the focus is on securing approvals, documenting compliance, and thoroughly assessing risks. This involves presenting a detailed model proposal outlining its purpose, methodology, data requirements, and potential benefits, ensuring alignment with the organization's risk appetite and governance framework. Key stakeholders must sign off on the proposal, attesting to its compliance with ethical and regulatory standards. A comprehensive risk assessment identifies potential financial, operational, reputational, or ethical risks associated with the model informing mitigation strategies. Transparent reporting mechanisms are established to track model performance, risks, and incidents, ensuring continuous monitoring and accountability.

Identify Inherent AI Risks and Define Preliminary Metrics

Definition, development, and implementation of key metrics and performance indicators for the AI Governance Framework and Reference Architecture, i.e., implement procedures to document key metrics and performance indicators on the effectiveness of your AI governance and model development lifecycle processes. Implement a crawl-walk-run approach to implementing a centralized AI governance model and Automation Center of Excellence (CoE). The following metrics can be measured within the Model Proposal and Feasibility stage:

Table 4-1. Predefined AI Metrics for Model Proposal

Metric Area	Metric Description
Problem Definition	• Business Value • Potential ROI (Return on Investment) • Cost Savings • Increased revenue potential • Alignment with strategic goals (quantifiable measures)
Bias Assessment	• Disparity in predictions across demographic or protected groups • Statistical measures of bias (e.g., disparate impact, demographic parity)
Performance Metrics	• Accuracy, Precision, Recall, F1-Score (classification) • Mean Absolute Error (MAE), Mean Squared Error (MSE), R-squared (regression) • Area Under the ROC Curve (AUC-ROC) • Consistency of performance across different datasets or over time. • Establish policies and processes that define impact/risk assessment scales for measuring the potential impact of the AI system. Red-amber-green (RAG) scales may be qualitative or entail simulations or econometric approaches.
Fairness Metrics	• Equalized Odds, Statistical Parity, Disparate Impact • Counterfactual Fairness
Explainability	• Importance of features contributing to predictions • Ability to generate human-understandable explanations (e.g., LIME, SHAP values)
Resource Usage	• Model size (number of parameters) • Computation time for training • Inference speed

CHAPTER 4   DESIGNING A WELL-GOVERNED AI LIFECYCLE MODEL

Model Development and Testing

Model development begins with feature engineering, transforming raw data into meaningful attributes, and selecting the most relevant to the problem. Next, carefully choose an algorithm suited to the task (classification, regression, etc.), balancing complexity with the need for explainability – consider techniques like decision trees or tools like LIME and SHAP for understanding complex models. Optimize the algorithm's configuration through hyperparameter tuning (grid search, random search, etc.), then split data for training, validation, and testing.

The training process involves iterative refinements, followed by performance and fairness assessments on the validation set, potentially requiring adjustments or mitigation techniques. Here is a breakdown of the detailed procedures during the model development and testing phase of the AI lifecycle:

Feature Engineering and Why It Matters

- Data transformation and creation of meaningful features from raw data.

- Feature selection to identify the most relevant features for the model.

- **Improved Model Performance** – The quality of your features often impacts model performance more than the algorithm itself. Relevant and informative features make it easier for AI models to learn patterns and make accurate predictions.

- **Addressing Data Complexity** – Real-world data is often messy and contains irrelevant, noisy, or redundant information. Feature engineering helps extract the most meaningful signals from this complex data.

CHAPTER 4   DESIGNING A WELL-GOVERNED AI LIFECYCLE MODEL

- **Reducing Model Size and Computational Cost** – Focusing on the most notable features can decrease model complexity, making it faster to train and deploy.

Key Feature Creation and Engineering Techniques

- **Feature Scaling and Normalization** → Rescaling features to a common range to improve algorithm convergence (e.g., standardization, min–max scaling).

- **Text Transformation** → Converting textual data into numerical representations for machine learning algorithms (e.g., Bag-of-Words, TF-IDF, word embeddings).

- **Feature Interaction** → Creating new features by combining existing ones and capturing nonlinear relationships.

- **Date/Time Manipulation** → Extracting components like day of the week or time of day from timestamps.

Feature Selection

- **Filter Methods** → Select features based on statistical measures like correlation with the target variable or variance.

- **Wrapper Methods** → Use a model to evaluate subsets of features, selecting those that improve performance.

- **Embedded Methods** → Build feature selection into the algorithm's training process (e.g., Lasso regression with regularization).

Domain Expertise is Crucial

While automated techniques exist, feature engineering often benefits from domain knowledge. Understanding your data and the problem you are solving helps guide the creation and selection of meaningful features. Feature engineering is an iterative process and rarely a one-and-done strategy. Experiment with different techniques and evaluate their impact on model performance through the development and testing cycle.

Algorithm Selection

Choose appropriate algorithms based on the problem type (classification, regression, clustering, etc.) and desired trade-offs between explainability and complexity. Consider using explainable AI techniques (e.g., decision trees, linear models) or tools for interpreting complex models (LIME, SHAP).

Hyperparameter Tuning

Optimize algorithm configuration through grid search, random search, or Bayesian optimization.

Training and Evaluation (Validation Set)

Split data into training, validation, and testing sets. Train the model on the training set, iteratively refining parameters and settings. Assess model performance on the validation set using relevant performance and fairness metrics. Adjust hyperparameters, explore different algorithms, or implement bias mitigation techniques as needed.

Testing and Independent Testing (Test Set and Scenario Testing)

Evaluate the final model on a holdout test set that was not used during training or validation. This provides an unbiased assessment of generalization capabilities. To assess robustness, evaluate model behavior under diverse scenarios, including edge cases and potential adversarial inputs.

Bias Assessment

Measure fairness metrics across protected groups or sensitive attributes. If biases are detected, apply fairness-aware algorithms or post-processing techniques.

Human-in-the-Loop (HITL)

HITL is a design principle incorporating human judgment, expertise, and oversight into AI decision-making processes. It creates a partnership between humans and AI systems, especially in high-risk domains where complete automation might be undesirable or unethical. Integrate human judgment and oversight as an additional validation layer for high-risk applications, especially for complex or uncertain predictions. HITL is used when your model makes complex or sensitive decisions and where AI models might lack the necessary context, nuance, or ethical reasoning to make decisions independently. When the need to address bias arises, humans can identify and correct biases the AI model might have learned from data or that emerge over time. HITL builds AI trust and fosters transparency and accountability, enhancing user trust in AI systems. As such, HITL manages uncertainty well; when AI model confidence is low or during unexpected scenarios, humans can provide guidance.

Interpretability Analysis

Use techniques like LIME, SHAP, or counterfactual explanations to understand how the model makes decisions. This aids in trust-building and debugging.

Documentation and Iteration

Recordkeeping → Thoroughly document algorithm choices, hyperparameters, preprocessing steps, performance metrics, and bias analyses for reproducibility and auditing. Model development is rarely linear. Expect to revise feature engineering, try different algorithms, and refine models based on testing results.

Defining HITL Triggers → Determine when the AI system requires human intervention. This could be based on

- **Confidence Thresholds** → Low certainty predictions routed to humans for review.

- **Specific Conditions** → Certain inputs, data anomalies, or high-impact cases trigger human intervention.

- **User Request** → Users can request a human review for any AI-generated output.

Levels of Intervention:

- **Human Assistance** → AI provides recommendations, but humans decide.

- **Human Consultation** → AI makes initial decisions, but humans can override if needed.

- **Active Learning** → Human judgments are used to retrain or refine the AI model, improving it over time.

- **Interface Design** → Develop clear and intuitive interfaces for human experts to

- **Review AI Recommendations** – Provide necessary context alongside the model output.

- **Explain Decisions** → Enable humans to provide rationales behind their interventions.

- **Provide Feedback** → Collect human feedback to improve the AI model.

Key Considerations

Use version control and ensure detailed documentation to prioritize reproducibility. Continuously monitor data quality and address any shifts in data distributions that might affect performance. Design tests that challenge the model's limits and assumptions through testing rigor.

Speed vs. Accuracy Trade-Off: HITL may introduce latency but can improve accuracy and address ethical concerns. Humans involved should have the necessary domain expertise to make informed judgments. Continuously incorporate human expertise to refine the AI model, creating a learning system or feedback loops. Finally, ensure audibility and log human interventions and rationales for traceability and analysis.

Examples of HITL

- **Medical Diagnosis** → AI assists with image analysis, but doctors provide final diagnoses and treatment plans. Content Moderation: AI flags potentially harmful content, with humans making final decisions.

- **Financial Decisions** → AI models support loan approvals, but underwriters review high-risk cases or exceptions.

Model Review and Validation

The AI model validation process rigorously assesses a model's performance, fairness, and robustness before deployment. This involves evaluating the model on a held-out dataset (not used during training) using a variety of performance metrics predefined and tailored to the task, alongside fairness metrics, to detect biases across different subgroups of data. AI Model Validation extends beyond accuracy, simulating real-world scenarios with diverse inputs to evaluate resilience against adversarial attacks, noise, and unexpected data shifts. Human experts may independently review results, and the process includes bias assessment and potential mitigation techniques to ensure the model aligns with ethical guidelines and responsible AI principles.

Why Independent Validation Matters

- **Bias Prevention** → A distinct perspective helps identify hidden biases or blind spots that the model developers might have missed.

- **Robustness Verification** → Independent validation rigorously assesses the model's performance under diverse and challenging scenarios.

- **Objectivity** → Builds trust in the model by having an unbiased third party evaluate its performance and fairness.

- **Regulatory Compliance** → Often required in industries like finance or healthcare.

Key Strategies for Model Validations and Validators

Rigorous Testing with Diverse Datasets → Go beyond standard accuracy metrics. Use multiple datasets reflecting real-world conditions, edge cases, and potential adversarial examples to proactively evaluate model performance and robustness.

Cross-Validation and Bootstrapping → Techniques like K-fold cross-validation or bootstrapping provide more reliable estimates of model performance on unseen data, reducing the risk of overfitting.

Benchmarking Against Baselines → Compare the AI model's performance to simpler statistical models and human judgment (where applicable). This helps gauge whether the model's complexity is justified by its improvement over simpler approaches.

Statistical Bias Detection Tools → Employ specialized toolkits or libraries to measure and quantify diverse types of bias (e.g., disparate impact, demographic parity) across various subgroups or sensitive features within your data.

Independent Validation → Involve an independent team with fresh perspectives for a rigorous audit of model performance, fairness, and compliance. This helps uncover blind spots and reduces internal biases.

Model Validators play diverse roles, including domain experts who provide context and identify ethical risks, data scientists who analyze performance and debug issues, and compliance specialists who ensure alignment with regulations and internal policies.

Table 4-2. Key Steps in Independent Model Validation

Model Validation Key Steps	Description
Define Scope and Criteria	• Outline the specific aspects of the model to be validated (performance, fairness, explainability, robustness). • Determine acceptable thresholds for performance metrics and fairness measures based on the use case's risk profile.
Establish Independent Team	• Form a team with the necessary technical and domain expertise, separate from the model development team. • Consider engaging external experts or auditors for specialized knowledge or regulatory compliance.
Secure Data Access	• Provide the validation team with a hold-out dataset never used in development or testing. • Prioritize security and privacy through access controls and potential anonymization techniques.

(continued)

Table 4-2. (*continued*)

Model Validation Key Steps	Description
Comprehensive Testing	• Evaluate performance metrics across all relevant subpopulations to detect bias. • Conduct stress testing with edge cases, adversarial examples, and changing input distributions. • Apply explainability methods to ensure the model's logic is sound.
Rigorous Documentation	• Thoroughly document all validation procedures, datasets, results, and any identified issues. • Establish clear reporting chains and escalation procedures for potential concerns.
Recommendations and Remediation	• Provide clear recommendations for model improvement, bias mitigation, or additional testing. • Track the implementation of remediation actions and re-validate as needed.
Data Quality	• The validation dataset should be representative of real-world scenarios and carefully checked for quality issues
Ensure Adequate Skills and Expertise	• The independent validation team should have strong data science, machine learning, and statistical analysis skills. • In high-stakes domains, consider partnering with a third-party auditing firm for additional scrutiny.
Communication and Collaboration	• Maintain open channels between the model development and validation teams to facilitate remediation and continuous improvement.

(*continued*)

Table 4-2. (*continued*)

Model Validation Key Steps	Description
Regular Validation Schedule	• Establish a cadence for independent model validations, especially in dynamic environments or evolving datasets.
Audit Trails	• Maintain auditable records of the validation process and decision rationales.

AI Model Approval

Here is a basic AI Model Approval process framework, including steps, stakeholders, and key considerations. Model Submission and Documentation – Your MLOps Engineers or Developers should focus on submitting a comprehensive model package, including

- Model code and artifacts
- Development and training datasets (with metadata)
- Performance metrics on validation sets
- Fairness and bias assessments
- Explainability reports (if applicable)
- Risk assessment outlining potential harms, use cases, and mitigation strategies

Multidisciplinary Review Team

- **Technical Experts** → Data scientists and ML engineers assess performance, reproducibility, and technical robustness.

- **Domain Experts** → Evaluate the model's suitability for the problem and identify potential errors or biases in outputs.

- **Compliance and Ethics Specialists** → Review alignment with regulations, data privacy policies, and the organization's ethical principles.

Approval Criteria

- **Performance Thresholds** → The model meets pre-defined accuracy, precision, recall, or other relevant metrics.

- **Fairness and Bias** → Bias measures are within acceptable ranges, or mitigation strategies are in place.

- **Ethical Considerations** → The model's potential impact aligns with organizational values and adequately addresses risks.

- **Explainability (If Required)** → Decision logic can be explained in a way stakeholders understand.

- **Documentation** → Clear and complete documentation exists for future reference and auditing.

Decision and Communication

- **Conditional Approval** → This may require model refinements, additional testing, or ongoing monitoring plans.

- **Rejection** → The model is sent back for development with apparent reasons and recommendations.

- **Approved** → The model is cleared for deployment, with a schedule for regular review and re-approval as needed.

- **Stakeholder Communication** → Decisions are communicated transparently to developers, business owners, and potentially end users.

Post-Deployment Monitoring

- **Performance Tracking** → Dashboards monitor key metrics for degradation or concept drift.

- **Feedback Loop** → Users can provide feedback on model outputs, triggering additional review if necessary.

- **Scheduled Re-approval** → Models are periodically re-evaluated as data distributions change or new risks emerge.

Additional Considerations

- **Version Control** → Track changes to models, data, and approval decisions for auditability.

- **Risk-Based Approach** → High-stakes models (healthcare, finance) may require more stringent approval processes and independent validation.

In summary, and I reiterate, the AI model approval process involves submitting a detailed model package that includes code, data, performance results, and potential risk assessments. A cross-functional team of technical, domain, and compliance experts reviews this package against predefined criteria for performance, fairness, ethical impact, and documentation. The model may be approved outright, conditionally approved pending modifications, or rejected with feedback.

CHAPTER 4   DESIGNING A WELL-GOVERNED AI LIFECYCLE MODEL

Figure 4-3. *AI Model Approval Process*

After deployment, ongoing monitoring, user feedback mechanisms, and scheduled re-approvals ensure the model remains effective, safe, and aligned with your organization's values and AI objectives. *Note: This process can also facilitate an approval stage gate workflow or process.*

AI Model Deployment and Implementation

The AI model deployment and implementation phase bridges the gap between development and real-world use. It involves setting up the required infrastructure (servers, cloud resources, etc.), integrating the model into workflows or applications, and designing intuitive user interfaces. This phase also includes establishing processes for managing updates and version control and ensuring data preprocessing remains coordinated with the model's needs. Continuous monitoring of performance, potential concept drift, and biases is vital to maintaining the model's effectiveness and addressing any emerging issues.

Key Considerations Before Deploying AI Models

Security and Access Controls → Protect model code, data pipelines, and the production environment from unauthorized access or attacks. Implement role-based permissions as needed.

Ethical Oversight → Ensure ongoing monitoring aligns with ethical principles, particularly in dynamic real-world settings where new biases may emerge.

Documentation → Thoroughly record deployment configurations, integration details, performance baselines, and monitoring procedures to aid troubleshooting and reproducibility.

Additionally, I offer the following process steps recommendations:

Infrastructure Preparation → Select and provision the necessary hardware (servers, GPUs, cloud instances) or leverage containerization solutions (e.g., Docker, Kubernetes) for scalability and portability.

Model Packaging and Deployment → Prepare the model for production, optimizing for inference speed and resource usage. Choose a deployment method based on usage patterns and latency requirements (API endpoint, embedded in an application, batch processing).

Integration → Seamlessly connect the model to existing business processes or software systems, ensuring smooth data input and output flows.

User Interface (UI) Development → Design intuitive interfaces that present AI outputs for applications with direct user interaction. Provide mechanisms for users to give feedback on model predictions.

Version Control and Rollback Strategies → Implement versioning systems to track model changes, iterations, and associated datasets. Establish procedures for rolling back to previous versions if performance degrades or issues arise.

Monitoring Dashboard → Create comprehensive dashboards visualizing real-time performance metrics, concept drift indicators, and fairness measures. Set up alerts to flag potential anomalies or biases.

Feedback Mechanisms → Develop channels for users and stakeholders to provide qualitative feedback on model performance, potentially influencing retraining or refinement.

Continuous Improvement Pipeline → Establish a process for retraining models based on performance monitoring, new data availability, or changing business requirements.

Common AI Model Deployment Challenges

Performance and Latency → Ensuring the model can manage real-world workloads and provide results within acceptable time frames. Real-world data may have distributions or noise unseen during training.

Scalability → Managing sudden increases in demand or traffic without compromising performance or inducing bottlenecks.

Data Compatibility and Pipelines → Ensuring data in production matches the format and quality the model was trained on. This involves robust data preprocessing and quality checks within the deployment pipeline.

Integration Complexity is the process of seamlessly fitting the AI model into existing systems, which may have legacy technologies or complex architectures.

Model Monitoring and Maintenance → Tracking model performance, drift, and potential bias over time can be resource-intensive, especially in dynamic environments.

Explainability and Trust → Providing clear explanations for model decisions in production can be difficult, especially for complex models. This hinders trust and adoption.

Resource Constraints and Costs: Deploying AI models in some contexts (e.g., edge devices) may face limitations in computational power, memory, and energy consumption.

Addressing AI Model Deployment Challenges

- **Model Optimization** → Use quantization, pruning, or knowledge distillation to reduce model size and boost inference speed.

- **Load Balancing and Caching** → Distribute requests efficiently across multiple servers or use caching to oversee peak demand.

- **CI/CD Pipelines** → Automate testing, deployment, and retraining processes for agility and faster iteration cycles.

- **MLOps Practices** → I recommend borrowing from DevOps principles; MLOps emphasizes robust version control, continuous monitoring, and team collaboration.

Table 4-3. AI Model Deployment Tools

AI Model Deployment Tools	Description
Cloud Platforms (AWS, Azure, GCP)	Offers scalable infrastructure, machine learning services, and pre-built APIs to simplify deployment and management.
Containerization (Docker, Kubernetes)	Packages models and dependencies for portability and consistency across environments.
Serverless Functions (AWS Lambda, Azure Functions)	It focuses on code logic with the infrastructure scaling handled automatically, making it well-suited for event-driven or sporadic model usage.
Edge Computing	Deploying models directly on devices (IoT sensors, smartphones) for low-latency, offline use cases often require model optimization techniques.
Specialized ML Platforms	Tools like TensorFlow Serving, MLflow, or Seldon Core provide standardized model serving, monitoring, and experiment tracking functionalities.

Ongoing Monitoring of AI Models

While emphasizing Continuous Improvement, you should also treat AI Model Monitoring as an ongoing and iterative process, not a one-time check. Regularly review monitoring strategies, adapt to evolving risks, and prioritize continuing communication between the AI team and stakeholders. I often see a common challenge: "What metrics and performance indicators should I track?"

Key Areas to Monitor

Performance Deterioration → Track accuracy, precision, recall, or other relevant metrics over time. Set alerts for significant declines that could impact business outcomes.

Data Distribution Shifts → Monitor the characteristics of incoming data for changes, unexpected values, or outliers. Sudden shifts could indicate concept drift, leading to inaccurate predictions.

Emerging Biases → Continuously assess fairness metrics across relevant subgroups. Track if model outputs disproportionately favor or disadvantage protected groups, indicating evolving biases.

User Feedback → Implement channels for users to provide feedback on model outputs. Look for patterns in reported errors, unexpected behavior, or ethical concerns users raise.

Resource Utilization → Monitor computational resource usage (CPU, memory, network) to identify potential bottlenecks impacting response times or scalability.

Key Metrics for Monitoring

Performance Metrics → Track accuracy, precision, recall, F1-score, or other appropriate measures relevant to the problem type.

Fairness Metrics → Monitor disparities across protected groups or sensitive attributes that could indicate emerging bias.

Data Distribution → Track input data and output distributions for changes or outliers, signaling potential concept drift.

Explainability (If Applicable) → Monitor shifts in feature importance or explanations provided by the model to identify instability.

User Feedback → Collect and analyze qualitative feedback from users to identify unexpected errors or biases.

Technical Implementation

Monitoring Dashboards → Visualize key metrics over time with threshold-based alerts for anomalies or performance degradation.

Data Quality Checks → Implement automated checks on input data for completeness, consistency, and conformity with expected formats.

Feedback Collection Mechanisms – Design easy-to-use channels for users to provide qualitative feedback and report issues.

Logging and Auditing → For debugging and compliance, maintain detailed logs of model predictions, inputs, performance data, and any interventions.

Go Beyond Technical: Processes and Responsibilities

Remediation Plans → Define procedures for handling declining performance, concept drift, or fairness issues. This may involve retraining, data updates, or human intervention.

Clear Ownership → Establish clear roles for those who monitor dashboards, analyze anomalies, and execute remediation actions.

Scheduled Reviews → Besides automated alerts, schedule regular reviews of model performance and fairness metrics.

Operational Considerations

Retraining Pipelines → Establish processes for retraining models when performance declines or new data becomes available.

Ownership – Define clear roles and responsibilities for those who actively monitor the dashboards, investigate issues, and take mitigation actions.

Communication Protocols → Establish procedures for alerting stakeholders about critical issues and communicating when models have been updated or retrained.

Summary and Thoughts

Designing a successful AI lifecycle model hinges on infusing governance from the beginning. Start by establishing clear ethical principles that guide all AI-related projects. Form a cross-functional governance committee involving stakeholders from data science, legal, risk management, and relevant business units. Define roles and responsibilities with a RACI matrix, ensuring accountability at every lifecycle stage.

CHAPTER 4   DESIGNING A WELL-GOVERNED AI LIFECYCLE MODEL

Embed robust governance practices throughout the AI lifecycle. Mandate data governance, quality control, and bias checks. Set standards for model development, testing, and ongoing monitoring. Implement explainability requirements based on model complexity and risk. Employ a risk assessment framework that proactively identifies potential harms like discrimination, security vulnerabilities, or operational failures. Establish an incident response plan for various AI-related issues and conduct regular audits to evaluate the effectiveness of your governance framework.

To achieve stated AI governance and risk management goals, define a clear set of metrics, KPIs (Key Performance Indicators), and KRIs (Key Risk Indicators) that align with your organization's specific priorities. These could include metrics on fairness, explainability, model drift detection, incident response times, compliance audit results, and stakeholder satisfaction with governance processes. Set realistic targets, track these metrics over time, and establish feedback loops. Regularly review results with your governance committee and use this data to identify areas for improvement and proactively address potential risks. Prioritize transparency by communicating KPIs to relevant teams, demonstrating progress toward a responsible and well-governed AI approach.

Quiz

1. What is the first step in designing a well-governed AI development lifecycle?

    A) Implementing rigorous testing

    B) Establishing clear stakeholder communication

    C) Beginning with careful problem identification

    D) Incorporating bias mitigation techniques

CHAPTER 4   DESIGNING A WELL-GOVERNED AI LIFECYCLE MODEL

2. Which of the following is NOT a focus during the AI model development stage?

   A) Balancing explainability with accuracy

   B) Incorporating bias mitigation techniques

   C) Maintaining thorough documentation

   D) Implementing continuous monitoring

3. During model deployment, which practice ensures traceability?

   A) Human oversight

   B) Version control

   C) Rigorous testing

   D) Regular audits

4. What should be prioritized to ensure responsible data collection practices?

   A) Quality standards and bias assessments

   B) Explainability and accuracy

   C) Testing and independent review

   D) Stakeholder communication

5. True or False: During model development, it is important to balance explainability with accuracy.

6. True or False: Continuous monitoring and feedback loops are only necessary before deploying the AI model.

7. True or False: A well-governed AI lifecycle model includes regular audits and incident response plans.

CHAPTER 4 DESIGNING A WELL-GOVERNED AI LIFECYCLE MODEL

8. To design a well-governed AI development lifecycle, begin with careful _____ identification.

9. During model development, balance _____ with accuracy, incorporate bias mitigation techniques, and maintain thorough documentation.

10. Promote clear _____ communication and foster a cross-functional collaborative team throughout the iterative process.

CHAPTER 5

Aligning AI Governance with Other Internal Governance Models for Trustworthy AI: "The Convergence of Governance Frameworks"

Enhancing and operationalizing automation processes and controls for trusted AI requires an effective method for understanding the business problem AI is attempting to solve and defining the underlying objectives of the governance and risk management framework. Once the underlying

CHAPTER 5   ALIGNING AI GOVERNANCE WITH OTHER INTERNAL GOVERNANCE MODELS FOR TRUSTWORTHY AI: "THE CONVERGENCE OF GOVERNANCE FRAMEWORKS"

objectives have been defined and socialized, you can identify common controls that can be shared across your business units or lines of business (LOBs). Based on my 25 years of experience in Governance, Risk, and Compliance management, an effective AI Governance strategy requires a deep and wide approach, capturing the adjacent and interoperable Governance Models and Processes within your organization. Hence, this chapter is based on the *"The Convergence of Governance Frameworks."*

The *following illustration* effectively aligns the multiple governance frameworks and models.

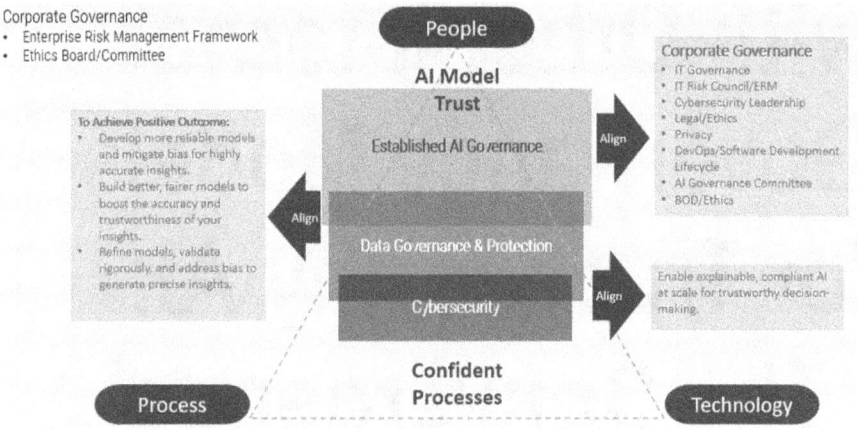

Figure 5-1. *Governance Alignments*

Establish AI Governance

The characteristics of trustworthy AI are integrated into organizational policies, processes, and procedures. These are central components of effective AI governance and risk management and fundamental to individual and organizational accountability.

CHAPTER 5   ALIGNING AI GOVERNANCE WITH OTHER INTERNAL GOVERNANCE MODELS FOR TRUSTWORTHY AI: "THE CONVERGENCE OF GOVERNANCE FRAMEWORKS"

AI Governance → AI Governance and Model Risk Management bring accountability and traceability to AI/ML models throughout their lifecycle. AI governance also involves ethical design, regulatory compliance, risk management, transparency, stakeholder engagement, and human oversight.

AI Ethics → AI ethics involves fairness, non-discrimination, transparency, accountability, privacy, security, and social benefit.

Data Governance → Data governance for AI involves data quality, security, privacy, bias mitigation, and data lineage.

AI Security Governance → AI security governance and risk management are essential for building trust and mitigating risks in AI systems.

This book highlights crucial aspects of responsible AI development and deployment. It emphasizes the need for robust AI security measures to protect systems and build trust. AI governance frameworks are essential, covering ethical design, compliance, risk management, and transparency. Model governance is vital for ensuring accountability across the AI lifecycle.

In addition, the book stresses the importance of AI ethics in promoting fairness, avoiding discrimination, and prioritizing social benefits.

Finally, it underscores that effective data governance, including data quality, privacy, and bias mitigation, is the foundation of responsible AI.

AI Ethics and Governance Strategy

Foundational Principles. Articulate and document your core values and define the ethical principles that guide your organization's AI endeavors (e.g., fairness, transparency, accountability, privacy). Establish an Ethics Committee and form a cross-disciplinary team with representation from technical, legal, compliance, and business stakeholders to advise on ethical considerations.

CHAPTER 5   ALIGNING AI GOVERNANCE WITH OTHER INTERNAL GOVERNANCE MODELS FOR TRUSTWORTHY AI: "THE CONVERGENCE OF GOVERNANCE FRAMEWORKS"

AI Ethics Framework Development. Start the process with an AI Ethics Risk Assessment: Develop a framework to identify and assess potential ethical risks throughout the AI lifecycle, from problem definition to data collection and model deployment. Create policies outlining responsible AI development practices, data usage guidelines, model validation procedures, and bias mitigation strategies.

- **AI Ethics and Responsible Use** → Fairness and Non-Discrimination: Ensure AI systems do not perpetuate or amplify existing biases, treating everyone equally regardless of protected characteristics (e.g., race, gender, disability). AI must treat everyone equally. Strive to implement proactive measures to identify and eliminate biases that can perpetuate societal inequalities.

- **Transparency and Explainability** → Design understandable AI systems to the extent needed. When decisions significantly impact, please explain how the AI reached them. People deserve to understand how AI decisions impact them. Explainability isn't optional; it's essential, especially for high-stakes decisions.

- **Accountability** → Clear lines of responsibility for AI systems' design, deployment, and outcomes. This includes having mechanisms for redressing if things go wrong. Clear ownership and responsibility for AI systems are crucial. If something goes wrong, there needs to be avenues for course correction. Your goal is to harness the power of AI to improve lives and address major societal problems.

- **Privacy and Security** → AI systems should protect user data and prevent unauthorized access or misuse. They should also respect privacy expectations and comply with regulations. Protecting user data and safeguarding systems is a cornerstone of trustworthy AI. Privacy must be designed, not just an afterthought.

- **Social Benefit** → Promote the use of AI for the good of society, addressing challenges like inequality, healthcare, and environmental issues.

- **Don't Assume Neutrality** → Acknowledge that AI has the potential for bias, harm, and unintended consequences. Approach AI development with a critical mindset, not assuming it is inherently fair or objective. Humans build AI and reflect our biases, both conscious and unconscious. Create unique and essential approaches to prevent harmful outcomes.

Operationalization and Tools

Algorithmic Bias Detection → Integrate bias detection toolkits and statistical techniques into the model development and testing phases.

Explainability Methods → Employ methods like LIME, SHAP, or counterfactual explanations to understand model decision-making and identify potential red flags.

Privacy-Preserving Technologies → Explore techniques like differential privacy or federated learning to protect sensitive data during model training and use.

Create a clear set of values and guidelines that align with your organization's mission. This should cover fairness, transparency, privacy, accountability, and social benefit. Implement a comprehensive AI ethics framework encompassing governance, diverse teams, bias mitigation, explainability, responsible data practices, impact assessments, audits, incident response, and continuous improvement based on feedback.

Capacity Building

Training and Education → Provide comprehensive training for employees on AI ethics principles, bias mitigation, and responsible data practices.

Up-Skilling Data Scientists → Offer specialized training in ethical AI development, emphasizing fairness-aware algorithms and explainability.

Transparency and Communication

Documentation → Thoroughly document AI system design choices, datasets, and decision rationale to enable auditing and accountability.

Plain-Language Explanations → Develop capabilities to provide clear explanations of AI model outputs for non-technical users, fostering trust.

Stakeholder Engagement → Maintain open communication channels with internal and external stakeholders to address concerns and incorporate feedback.

Continuous Auditing and Improvement

Independent Audits → Schedule regular audits by independent experts or third-party organizations to assess ethical compliance and identify potential shortcomings.

Feedback Mechanisms → Collect feedback from end users and those impacted by the AI systems to detect unintended consequences or biases.

Iterative Refinement → Treat AI governance as an ongoing process. Adapt policies and practices based on learnings, new risks, and evolving regulations.

Key Considerations

Context Matters → Tailor this strategy to your organization's size, specific AI use cases, and regulatory environment.

Collaboration Is Key → Foster collaboration between technical teams, legal experts, domain specialists, and ethicists to address AI challenges holistically.

Strong Governance → Establish clear roles, committees, and processes dedicated to overseeing AI development and use, ensuring ethical principles are embedded throughout.

CHAPTER 5 ALIGNING AI GOVERNANCE WITH OTHER INTERNAL GOVERNANCE MODELS FOR TRUSTWORTHY AI: "THE CONVERGENCE OF GOVERNANCE FRAMEWORKS"

Diverse Teams, Diverse Thinking → Actively build teams with varied perspectives to help uncover potential biases and blind spots during development.

Proactive Bias Management → Implement rigorous bias testing throughout the AI lifecycle, with ongoing mitigation strategies and correction plans.

Explainability As a Priority → When possible, use AI models that offer insight into their decision-making process.

Foundation of Responsible Data → Emphasize ethical data collection, storage, and use. Comply with regulations and go beyond, respecting user privacy expectations.

Assess Impact Before Action → Critically evaluate the potential social and ethical consequences of AI systems before they are deployed.

Audits for Accountability → Regularly audit AI systems for fairness, accuracy, and emergent risks. Ensure transparency and accountability.

Respond and Rectify → Have a robust incident response plan for addressing AI-related harms or ethical breaches.

Feedback Drives Improvement → Create open channels for stakeholders to provide input on AI systems, driving continuous improvement and addressing concerns.

In summary, AI has incredible potential but also poses ethical risks. These ethical principles provide a framework to ensure the development and use of AI benefits all societies, not just a select few. Fairness ensures equal opportunity regardless of background. Transparency increases trust and helps detect problems early. Accountability clarifies who is responsible when issues arise. AI systems must respect user privacy and be secure. Finally, we should actively seek ways to use AI for social good, solving real-world problems responsibly.

CHAPTER 5   ALIGNING AI GOVERNANCE WITH OTHER INTERNAL GOVERNANCE MODELS FOR TRUSTWORTHY AI: "THE CONVERGENCE OF GOVERNANCE FRAMEWORKS"

Data Governance

AI data governance centers on ensuring the data fueling your AI systems are high-quality, secure, and compliant with both privacy regulations and ethical principles. It involves establishing robust data quality standards, implementing measures to protect data throughout its lifecycle, proactively addressing potential biases embedded in datasets, and tracking data lineage for auditability and reproducibility. Effective data governance lays the foundation for building fair, accountable, and reliable AI applications.

Data Governance Foundational Principles and Framework

Data Governance Framework → Develop a clear framework outlining your data governance policies, procedures, roles, and responsibilities. Ensure it aligns with existing data governance and protection initiatives and applicable regulatory requirements. Compliance regulations like GDPR often require demonstrating the origin and uses of personal data. Data lineage will provide the auditable trail for this.

Your data governance strategy and guiding principles must be clear, actionable, and adaptable to your organization's specific needs. The organization's thoughts on data must be well-aligned, and your data must be recognized as a strategic asset. In the constructs of AI, data should be treated as a valuable organizational asset, rationalized, and managed with the same rigor as financial records or human capital.

Data Quality and Integrity → I recommend prioritizing the accuracy, completeness, consistency, and timeliness of data across its lifecycle. Data quality standards and policies are well-defined, monitored, and enforced.

Data Accessibility (with Controls) → Your organization should promote the discoverability and accessibility of data for authorized users while implementing strict access controls to protect sensitive information.

Data Accountability and Ownership → Data stewardship roles are clearly defined, with individuals or teams responsible for data quality, compliance, and usage within their domains.

CHAPTER 5 ALIGNING AI GOVERNANCE WITH OTHER INTERNAL GOVERNANCE MODELS FOR TRUSTWORTHY AI: "THE CONVERGENCE OF GOVERNANCE FRAMEWORKS"

Compliance and Ethical Use → You must adhere to all relevant data privacy regulations (GDPR, CCPA, etc.) and prioritize the ethical collection, processing, and use of data.

Transparency and Communication → Maintain open communication channels about data governance policies, procedures, and decisions, fostering organizational trust.

Security and Risk Management → Data security is a core pillar. Implement robust and scalable measures to protect data against unauthorized access, exfiltration, breaches, and misuse.

Data Lineage and Provenance → Data lineage should encompass the complete historical record of a piece of data – where it originated, the transformations it underwent, how it moved between systems, and its ultimate usage in different AI models or analyses. Your provenance strategy should augment lineage with the records of processes, people (roles), and entities involved in this journey. Knowing a data point's lineage allows you to assess its reliability and fitness for your model's purpose, fostering increased trust and reproducibility. You can trace the lineage back to understand data quality issues and potential biases introduced during its journey or for reproducing model results. Imagine your AI model exhibits unexpected behavior, and the data used for training is drifting. Tracking data lineage can help pinpoint if the issue stems from errors introduced during data manipulation or biases ingrained in its source. Think of lineage and provenance as a "forensics tool."

The Core Values of Data Governance should focus on the ethical principles guiding your AI data governance strategy and highlight model fairness, transparency, accountability, privacy, and explainability.

How to Use These Principles

Foundation for Framework → These principles inform the development of your detailed data governance framework, ensuring alignment with your organization's values.

CHAPTER 5 ALIGNING AI GOVERNANCE WITH OTHER INTERNAL GOVERNANCE MODELS FOR TRUSTWORTHY AI: "THE CONVERGENCE OF GOVERNANCE FRAMEWORKS"

Decision-Making Guide → Refer to these principles when faced with data management choices, ensuring consistency and promoting ethical AI development.

Training and Education → Use these principles to communicate the importance of data governance practices to employees across the organization.

Bias Awareness and Mitigation → To promote fairness in and across AI systems, proactively assess and mitigate potential biases throughout the data lifecycle, from collection to model development.

Governance Committee → Establish a cross-functional governance committee composed of data scientists, legal experts, business stakeholders, domain representatives, and ethicists.

Data Quality Impacts Model Quality

Garbage in, Garbage Out. Errors, inconsistencies, or missing values in your training data will directly lead to poor model performance, regardless of how sophisticated the algorithm is. Data Cleaning, preprocessing, outlier detection, and enumerating missing entries are essential for building reliable AI models. Data relevance ensures that the data being used for training closely reflects real-world conditions where the model will be deployed. Mismatches in distribution can significantly undermine model accuracy.

Addressing Bias in Training Data

Representative Datasets → Aim to collect data that reflects the diversity of populations your model will encounter. Unbalanced datasets lead to models that perpetuate biases.

Bias Detection → Employ statistical tools to quantify bias and protected attributes (e.g., race, gender).

Mitigation Techniques → Apply rebalancing, synthetic data generation, or fairness-aware algorithms to reduce bias in training data.

Ensuring Reproducibility and Auditability

Data Versioning → Track the dataset versions used to train each model iteration. This is critical if you need to assess performance changes or address issues in production.

Detailed Metadata → Capture information about the source, preprocessing steps, bias assessments, and any known limitations of your training data.

Security and Compliance → Protect training data in line with privacy regulations and access controls – track data usage for auditability.

Additional Considerations

Data Augmentation and Governance extends to how you generate synthetic data or apply transformations. Ensure these processes are documented and don't introduce unintended biases. If your models involve human-labeled training data, be aware that human annotators can also introduce biases. Implement quality checks and multiple annotators where feasible.

Data Governance As Key to Responsible AI

Treating data governance as a core part of your AI development process helps build accurate, fair, and explainable AI models. It aligns with ethical principles and strengthens trust amongst users of your AI systems. Align AI data governance with your overarching data governance policies, AI-specific ethical guidelines, and evolving regulations.

Data Management and Lifecycle Governance

Data Inventory → Create and continuously update a centralized catalog of all data assets relevant to AI, including their sources, ownership, sensitivity levels, and lineage.

Data Quality Standards → Establish rigorous standards for data accuracy, completeness, consistency, relevancy, and timeliness across the entire AI data lifecycle.

Access Control → Implement role-based access control mechanisms, encryption, and secure transfer protocols to protect data at rest and in motion.

Data Lifecycle Management → Define policies for data retention, archival, transformation, and secure deletion, aligning with regulatory requirements and minimizing risks.

Model Governance Integration

Model Documentation → Mandate thorough documentation of data sources, preprocessing steps, model versions, performance metrics, and bias assessments.

Metadata Standards → Define clear metadata requirements for models and their associated data to ensure auditability and reproducibility.

Model Risk Management → Implement processes to assess the risks associated with AI models, including potential biases, unintended consequences, and vulnerabilities.

Data Privacy and Security

Compliance Mapping → Map relevant regulations (GDPR, CCPA, etc.) to your AI data practices and update your governance framework to incorporate their requirements.

Consent Mechanisms → Ensure appropriate consent mechanisms for collecting personal data, especially for high-risk or sensitive AI applications.

Privacy-Preserving Techniques → Explore techniques like differential privacy, anonymization, and federated learning to enable model training while protecting individual privacy.

Communication and Training

Clear Communication Channels → Promote open communication across teams about data governance policies and procedures.

Tailored Training → Provide regular training to all relevant personnel on data governance principles, ethical AI practices, and specific compliance requirements.

Continuous Data Governance Improvement

Audits and Monitoring → Schedule regular audits of AI data practices, models, and governance processes. Implement monitoring dashboards for key metrics and indicators.

Feedback Loops → Establish mechanisms to collect feedback from users, stakeholders, and those affected by your AI systems. Use this to refine processes and mitigate biases.

Evolution with Technology → Treat data governance as an ongoing endeavor. Adapt your strategy as advanced technologies, regulations, and ethical concerns emerge.

Bias Mitigation Techniques

Pre-processing. Rebalancing or Resampling → Adjust the distribution of your dataset to represent minority classes or underrepresented groups better. I recommend data augmentation practices to generate synthetic data samples for underrepresented groups to improve balance, preventing bias-trained data.

In-Processing. Create fairness-aware algorithms and employ algorithms that explicitly incorporate fairness constraints during the training process. Adversarial Debiasing is highly encouraged to train secondary models to identify bias in the primary model's predictions, helping the original model learn to counteract it.

Post-processing and Threshold Adjustments → Modify decision thresholds for distinct groups to promote equalized outcomes and introduce reject options or introduce a "reject" category for uncertain predictions, allowing for human review in potentially biased cases.

Metadata Management

Predefined Metadata Standards should be used to define clear schemas for essential metadata elements, including

- Data source and collection methods
- Preprocessing steps and transformations
- Known biases and limitations
- Quality metrics (accuracy, completeness)
- Privacy considerations and access controls

CHAPTER 5   ALIGNING AI GOVERNANCE WITH OTHER INTERNAL GOVERNANCE MODELS FOR TRUSTWORTHY AI: "THE CONVERGENCE OF GOVERNANCE FRAMEWORKS"

Metadata tools and data catalogs should be used to store and search metadata. Additionally, automated metadata capture tools streamline the documentation process. Figure 5-2 below highlights the importance of the data fabric and data acquisition layers.

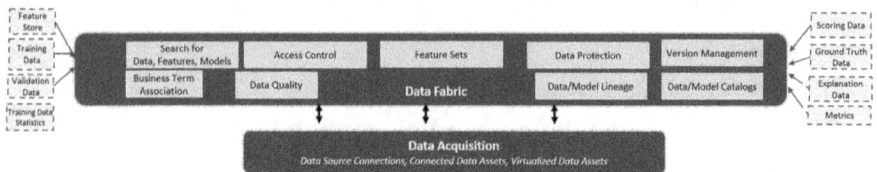

Figure 5-2. *Data Fabric and Acquisition Layer*

Let's take a detailed view of the components of the Data Fabric and Acquisition Layers. As part of defining your data needs and where you acquire your data from or its sources, I recommend "Problem Framing" or clearly explaining the problem your AI system aims to solve in conjunction with its data and data sources. This, in turn, dictates the type and format of data needed to train your models and will provide insights into the specific features or data points essential for training your model. Consider factors like model type, accuracy requirements, and potential biases. Explore existing data sources within your organization that might be relevant. This could include customer databases, sensor readings, or historical transaction records. Depending on your needs, you might consider purchasing data from reputable vendors, collaborating with other organizations, or leveraging publicly available datasets or other large language or foundational models.

Data Acquisition Methods. The method depends on the data source. This could involve setting up sensors, scraping public APIs, conducting surveys, or purchasing access to specific datasets. If your model requires labeled data (e.g., identifying objects in images), establish a labeling process with clear guidelines and quality control measures. Other data sources may include virtualized data assets or connected data assets.

CHAPTER 5   ALIGNING AI GOVERNANCE WITH OTHER INTERNAL GOVERNANCE MODELS FOR TRUSTWORTHY AI: "THE CONVERGENCE OF GOVERNANCE FRAMEWORKS"

By following a structured approach to data acquisition, you can ensure your AI models are built on high-quality, reliable data, leading to better performance and more trustworthy outcomes.

You will have many components to manage good-quality data within the Data Fabric Layer. For example, many of your data governance and protection controls will be managed within this layer, i.e., access management, data rationale and quality, data lineage, data catalogs, and data searches/indexing will be performed at this layer. **From a feature-set perspective**, the quality and relevance of your chosen features directly impact the model's ability to learn from its data to make accurate predictions. Including irrelevant features can confuse the model, while crucial features left out can limit its performance. Additionally, unique identifiers or version numbers to different iterations of a machine learning model can be managed within the Data Fabric Layer. The list of other data features that can be handled is outlined below:

- **Data Scoring** refers to applying a trained machine learning model to new data to generate predictions, classifications, or insights.

- **Ground Truth Data** → Ground truth data represents the absolute for a given task. It is the desired output that you want your AI model to learn to predict or classify.

- **Explanation Data** → It helps build trust by allowing users to understand the rationale behind a model's decision.

- **AI Metrics** → Measurements are used to gauge AI models' performance, fairness, explainability, and other aspects.

- **Training-Data Statistics** → The dataset that your AI model will learn from.

- **AI Validation Data** → Validation data is a separate dataset held out from the data used during model training. It aims to represent the real-world data your model will encounter when deployed.

- **AI Training Data** → The quality and quantity of your training data directly determine the capabilities and limits of your AI model

- **AI Feature Store** → A feature store is the sole source of truth for all the engineered features used to train and power machine learning models.

Unified Compliance Considerations

Mapping regulations to data practices and standards is a great tool for better understanding how regulations like GDPR or CCPA impact data collection, storage, usage, and deletion for AI-based data governance purposes. In addition, transparent and automated processes for collecting personal data, especially for sensitive use cases, should be implemented.

Explainability or right to explanation is key to data privacy and compliance. If required by regulations or ethical principles, I recommend a hybrid approach to model explainability and transparency necessary for AI-based decisions.

Privacy-Preserving Techniques → Explore differential privacy or federated learning to protect data while enabling model training based on AI model sensitivity and high-risk AI applications.

Customize these principles to reflect your industry, regulatory environment, and the specific data types used to train your AI models.

Key Data Governance Takeaways

- The quality of data is not a separate concern from the AI itself.

- Flawed or inappropriate data will sabotage even the most sophisticated algorithm.

- Emphasize that investing in data quality is investing directly in the success of their AI initiatives.

- Emphasize the direct link between AI success and the quality and quantity of data it can access.

- Highlight how advancements in AI are often fueled by the increased availability and management of data.

- Underscore the importance of balancing AI innovation with responsible data practices, ensuring privacy, and avoiding harmful biases.

- This strategy prioritizes ethical considerations and continuous improvement through feedback mechanisms and focuses on aligning with regulatory requirements and core values.

Specific Data Governance Implementation Steps

- **Data Quality Assessment** – The following sample outlines a detailed data quality assessment report focusing on open-source tools and techniques. The sample code (use with caution) primarily centers around Pandas Profiling for its rich insights.

- **Metadata Management** – Discuss best practices for metadata schemas, recommend tools (e.g., data catalogs), and explore how they link to model governance.

- **AI System Auditing** – Design an audit plan that includes data lineage, model decision rationale, bias metrics, and remediation steps when issues are found.

- **Install Libraries** – Ensure you have Pandas and Pandas Profiling installed. A simple pip installs pandas-profiling, which should suffice.

- **Import Data** – Load your dataset into a Pandas DataFrame for easy manipulation.

Generate Profile Report (Pandas Profiling)

- **Profile Creation** – A few lines of code generate a comprehensive report:

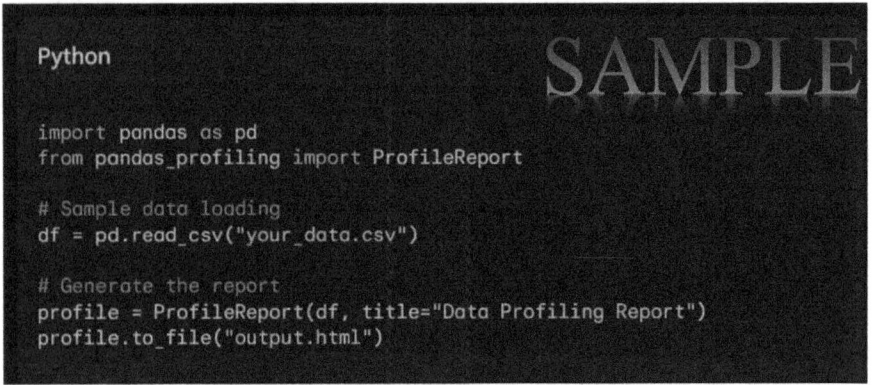

Figure 5-3. Sample Python Data Profiling Report

Report Analysis: The generated HTML report offers these sections:

- **Overview** – General dataset statistics (number of rows, columns, data types, etc.)

- **Variables** – Detailed analysis of each feature

- **Numerical Features** – Distributions, histograms, quantiles, and outliers

- **Categorical Features** – Category counts, bar charts.

- **Interactions** – Highlights pairwise correlations between features.
- **Missing Values** – Counts and analysis of missing data patterns.

Identify Quality Issues and Inconsistencies

- Unexpected data types (e.g., strings in a numeric column)
- Format errors (e.g., inconsistent timestamps)
- Incoherent values (e.g., negative ages)

Outliers (the following outliers will be identified)

- Statistical methods within the report help flag potential outliers.
- Visualizations like box plots aid in outlier examination.
- Domain knowledge for what constitutes a reasonable range is key.

Missing Values

- Percentage of missingness per feature
- Patterns of missingness (random, clustered, or systematic)

Data Cleansing and Preprocessing

- Handling Incompatibilities and data type conversions where needed
- String cleaning (e.g., fixing date formats)

CHAPTER 5   ALIGNING AI GOVERNANCE WITH OTHER INTERNAL GOVERNANCE MODELS FOR
           TRUSTWORTHY AI: "THE CONVERGENCE OF GOVERNANCE FRAMEWORKS"

Outlier Treatment

- Removal if they are erroneous.
- Capping/Flooring to retain data but limit impact.
- Domain expertise is critical for deciding the best approach.

Missing Value Addressing

- Remove rows/columns with excessive missing data.
- Imputation (e.g., mean, median) if appropriate. Techniques specific to feature type/model used.

Additional Open-Source Tools for Bias and Privacy Preservation

- **AIF360 (IBM)** → Open-source toolkit with various bias detection algorithms.
- **Fairlearn (Microsoft)** → Focuses on fairness assessment and mitigation strategies.
- **Specialized Tools** → There are emerging tools tailored to detecting bias in image datasets, text data (NLP), and more.
- **Differential Privacy** → Methods for adding calculated noise to datasets provide statistical privacy guarantees.
- **Federated Learning** → Trains models on decentralized data without sharing raw data.
- **Homomorphic Encryption** → Allows computations on encrypted data, useful for secure model inference.
- **NumPy** → Powerful array computation for detecting outliers and finding NaN (Not a Number) values.

- **Visualization Libraries** → (Matplotlib, Seaborn): Customize plots for deeper exploration.

- **Important Considerations** → Understanding the data's context is vital to interpreting these outliers or patterns.

- **Iterative Process** → Data quality checks should be incorporated throughout the AI development lifecycle.

In summary:

Make Sure Your AI Models Get the Right Data: A Business Perspective

Like any high-performing team, AI models need the right data to function at their best. Data governance and quality management are the secret sauce, ensuring your models are fed the right information at the right time.

Data Quality: The Foundation of Good Decisions

Data profiling continuously examines data to ensure it is complete (no missing values), consistent (formatted correctly), and follows expected patterns. This ensures errors or inconsistencies do not mislead the model.

Data Cleaning and Transformation

Imagine this as preparing the data for the model. We fix errors, standardize formats (i.e., converting currencies), and may even create new features from existing ones (i.e., a customer's total purchase history). This ensures the data is in a usable format and provides the most relevant information for the model.

Data Monitoring

I strongly recommend continuously checking data quality. This involves setting up alerts to catch anomalies (unusual patterns) that might indicate data issues or changing trends. Early detection prevents feeding the model with bad data, leading to potentially poor decisions.

Data Management: Keeping Track of Your Most Valuable Asset

Datasets are the core data collections used to train and run your AI models. Data governance ensures clear descriptions of each dataset, including its purpose, version history, who owns it, and access restrictions. This transparency is crucial for maintaining data integrity and responsible use.

Metadata: Think of it as "data about data."

Metadata provides critical details on how the data was collected, what it represents, and how it relates to other data points. This rich information helps track the data's lineage (where it came from), allows for model reproducibility (rebuilding the model with the same data), and ensures compliance with regulations.

Data governance and quality management are essential for building trust and ensuring the effectiveness of your AI initiatives. Robust AI data governance is crucial for ensuring data quality, security, and ethical use across the AI lifecycle. This directly translates into improved AI model performance, reduced risks associated with security breaches and biases, enhanced stakeholder trust, streamlined operations, and competitive advantage. Conversely, neglecting AI data governance leaves organizations vulnerable to inaccurate and biased AI models, reputational damage due to privacy violations or unethical outcomes, regulatory penalties, and missed opportunities for innovation.

IT and Cybersecurity Governance

What Is IT Governance 101?

IT governance is a framework that ensures an organization's information technology (IT) investments, resources, and processes align with its business goals and strategies. It encompasses decision-making structures, accountability mechanisms, policies, and standards aimed at maximizing the value IT delivers while mitigating risks associated with its

CHAPTER 5 ALIGNING AI GOVERNANCE WITH OTHER INTERNAL GOVERNANCE MODELS FOR TRUSTWORTHY AI: "THE CONVERGENCE OF GOVERNANCE FRAMEWORKS"

use. IT governance involves a partnership between business executives, IT leadership, and stakeholders to ensure technology investments are responsible and transparent and contribute to the organization's overall success.

IT governance rose to prominence in the late 1990s and early 2000s due to increasing business reliance on technology. High-profile IT failures and scandals underscored the risks of poor management, while frameworks like COBIT and ITIL offered solutions. Regulatory pressures, like the Sarbanes-Oxley Act, further emphasized IT controls and accountability, making IT governance a top priority for organizations seeking to manage technology risks and effectively align technology with business strategy.

By now, I'm sure you're asking, "What does IT governance have to do with AI governance and model risk management?"

IT governance plays a foundational role in AI governance and model risk management. It provides the overarching principles and processes guiding the responsible use of technology resources. This includes establishing data governance standards to ensure the quality and ethical collection of data used to train AI models. It also mandates IT risk management practices, including security controls to protect AI models and sensitive data. Moreover, it fosters transparency and accountability, requiring clear documentation of AI model development and decision-making processes. Ultimately, IT governance sets the stage for responsible AI development and deployment, ensuring AI initiatives comply with regulations and promote trust.

Typically, the Chief Information Officer (CIO) holds primary responsibility for IT governance within an organization. Here is the reason:

The *Chief Information Officer (CIO)* is primarily responsible for IT governance, ensuring technology aligns with business goals while managing risks and allocating resources efficiently. However, successfully implementing IT governance demands collaboration across the organization. The Board of Directors provides oversight and sets

CHAPTER 5   ALIGNING AI GOVERNANCE WITH OTHER INTERNAL GOVERNANCE MODELS FOR TRUSTWORTHY AI: "THE CONVERGENCE OF GOVERNANCE FRAMEWORKS"

risk tolerance, business leaders guide IT solutions to meet their unit's needs, and compliance officers ensure systems adhere to regulatory requirements. Effective implementation requires collaboration and shared accountability across

- Board of Directors, who provides oversight, setting the organization's risk tolerance and ensuring IT investments align with strategic goals.

- Business leaders who participate in IT decision-making ensure that IT solutions meet business needs.

- Compliance Officers, who collaborate with the CIO to interpret regulations and embed compliance requirements within IT systems.

What I Cybersecurity Governance 101?

Cybersecurity governance is a subset of IT governance that focuses on establishing a framework for managing an organization's cybersecurity risks. It involves defining clear policies, standards, and decision-making structures regarding the confidentiality, integrity, and availability of information assets. Cybersecurity governance aims to align security initiatives with business objectives, ensure compliance with regulations, and foster a culture of security awareness throughout the organization. This often includes implementing risk assessments, security controls, incident response plans, and ongoing monitoring.

Like IT governance, cybersecurity governance emerged as a major concern in the 1990s and early 2000s due to the rise of the Internet, which exposed organizations to new cyber threats. Alongside a rapid increase in the volume and value of digital data, high-profile security breaches exposed the risks of inadequate security. This confluence of events and the introduction of stricter data security regulations propelled security governance to the forefront of business priorities.

CHAPTER 5   ALIGNING AI GOVERNANCE WITH OTHER INTERNAL GOVERNANCE MODELS FOR TRUSTWORTHY AI: "THE CONVERGENCE OF GOVERNANCE FRAMEWORKS"

I'm sure you're also wondering, "What does cybersecurity governance have to do with AI governance and model risk management?"

Cybersecurity governance is foundational in AI governance and risk management because AI models and their associated datasets are highly valuable and vulnerable assets. Cybersecurity governance establishes policies and safeguards to protect the confidentiality, integrity, and availability of AI models, preventing unauthorized access, theft, or manipulation. It addresses issues like data encryption, access controls, and secure development practices to mitigate the risk of breaches that could compromise sensitive data or lead to biased or unreliable AI outputs. Additionally, cybersecurity governance frameworks guide incident response plans (*we will touch on later*) to quickly detect and address security threats targeting AI systems, minimizing potential damage, and ensuring continuity.

While several roles contribute, the Chief Information Security Officer (CISO) often plays a pivotal role in cybersecurity governance. This is because CISOs possess the deep technical expertise to understand cybersecurity risks, implement mitigation strategies, and oversee security operations. They work closely with business leaders to align security measures with strategic goals and ensure compliance with regulations. CISOs are responsible for developing security policies and incident response plans and fostering a security-conscious culture across the organization. They are accountable for maintaining a robust cybersecurity posture and safeguarding the organization's AI system assets.

...and how do we bring it all together to construct an effective AI governance framework?

Effective IT and cybersecurity governance form the bedrock of successful AI governance and risk management. This integrated approach focuses on establishing a framework of policies, processes, and controls to ensure the responsible use of technology and data throughout the AI lifecycle. Here is how it translates into action:

CHAPTER 5 ALIGNING AI GOVERNANCE WITH OTHER INTERNAL GOVERNANCE MODELS FOR TRUSTWORTHY AI: "THE CONVERGENCE OF GOVERNANCE FRAMEWORKS"

Firstly, IT and cybersecurity governance dictate the selection and implementation of data governance standards. This ensures the data used to train AI models is high-quality, unbiased, and collected ethically. Comprehensive data access controls and encryption measures, mandated by cybersecurity frameworks, safeguard sensitive data and prevent unauthorized access. Secondly, cybersecurity governance plays a critical role in protecting AI models themselves. Secure coding practices, vulnerability assessments, and penetration testing, all facilitated by strong cybersecurity governance, mitigate the risk of cyberattacks that could compromise AI models or manipulate their outputs.

Furthermore, IT governance fosters transparency and accountability within the AI development process. This involves documenting model development steps, decision-making rationale, and potential biases identified during training. This level of transparency, encouraged by IT governance, empowers stakeholders to assess the trustworthiness of AI models and identify potential risks.

Finally, IT and cybersecurity governance work together to implement robust incident-response plans for AI systems. These plans, guided by IT governance and cybersecurity frameworks, ensure any security breaches or malfunctions are swiftly detected, contained, and remediated. This minimizes potential damage to the organization's reputation, data integrity, and the reliability of AI outputs. Effective IT and cybersecurity governance is the foundation for building and deploying trustworthy AI. By ensuring data integrity, model security, and transparent development processes, these intertwined disciplines empower organizations to mitigate risks associated with AI and confidently leverage its full potential.

As we touched on incident response in the context of AI governance and risk management, additional cybersecurity core control families play a crucial role in preventing adversarial AI attacks and risks. Here is a list of cybersecurity core control procedures, along with the rationale for securing AI models:

CHAPTER 5 ALIGNING AI GOVERNANCE WITH OTHER INTERNAL GOVERNANCE MODELS FOR TRUSTWORTHY AI: "THE CONVERGENCE OF GOVERNANCE FRAMEWORKS"

Data Security

- Data classification (identifying sensitive data)
- Encryption at rest and in transit
- Strict access controls and authorization
- Data anonymization and de-identification techniques where appropriate

Data Classification → Identifying and classifying sensitive data used to train or interact with AI models is crucial. This allows for appropriate security measures based on the data's sensitivity level. For example, medical records or financial data require stricter controls than weather data.

Encryption → Encrypting data at rest (stored on servers) and in transit (traveling over networks) safeguards it from unauthorized access, even if intercepted by attackers. This protects individuals' privacy and prevents sensitive information from being used to manipulate AI models.

Access Controls and Authorization → Strict access controls dictate who can access data based on their role and responsibilities. This principle of least privilege minimizes the risk of unauthorized users tampering with data or compromising its integrity. Multi-factor authentication strengthens access control by requiring additional verification steps beyond a username and password.

Data Anonymization and De-identification → In certain scenarios, anonymizing data by removing personally identifiable information (PII) can enhance data privacy and mitigate ethical concerns associated with AI bias. De-identification techniques should also be employed, where some data elements are altered while preserving the data's utility for AI training.

Network Security

- Firewalls and intrusion detection/prevention systems (IDS/IPS)
- Network segmentation to isolate AI systems and sensitive data
- Secure remote access solutions (VPNs, etc.)

Firewalls and Intrusion Detection/Prevention Systems (IDS/IPS)
→ Firewalls function as gateways, filtering incoming and outgoing traffic based on pre-defined security rules. This helps block malicious traffic from reaching AI systems and the data they rely on. IDS/IPS systems continuously monitor network activity for suspicious behavior, detecting and potentially preventing cyberattacks aimed at exploiting vulnerabilities.

Network Segmentation → Dividing the network into smaller, isolated segments creates barriers between AI systems, sensitive data, and other parts of the network. This "compartmentalization" approach limits the potential damage if a breach occurs in one segment, preventing attackers from easily accessing critical AI resources.

Secure Remote Access Solutions → When remote access to AI systems is necessary, secure solutions like virtual private networks (VPNs) should be used. VPNs create encrypted tunnels, ensuring all data transmitted between the remote user and the AI system remains confidential and protected from interception.

System Security

- Vulnerability scanning and patch management
- Configuration hardening (removing unnecessary services, secure defaults)
- Endpoint protection (anti-virus, application whitelisting)

Vulnerability Scanning and Patch Management → Regularly scanning AI systems for vulnerabilities and promptly applying security patches is essential. This practice minimizes the risk of attackers exploiting known weaknesses in the system's software or configuration.

Configuration Hardening → This involves disabling unnecessary services and functionalities on AI systems, reducing the attack surface and potential entry points for malicious actors. Enforcing secure defaults for system configurations further minimizes the risk of misconfiguration leading to vulnerabilities.

Endpoint Protection → Implementing endpoint protection solutions like anti-virus software and application whitelisting safeguards AI systems from malware and unauthorized applications that could compromise their integrity or manipulate data used for training.

Identity and Access Management (IAM)

- Strong authentication (multi-factor authentication)
- Role-based access controls (RBAC) to enforce the principle of least privilege
- Regular user access reviews and auditing

Strong Authentication → Multi-factor authentication (MFA) goes beyond traditional username and password logins. It requires additional verification steps, such as a code sent to a mobile device, to ensure that only authorized users can access AI systems and sensitive data.

Role-Based Access Control (RBAC) → This principal grants access to data and functionalities based on a user's role within the organization. For example, data scientists might have access to training data, while business analysts might only have access to model outputs. RBAC prevents unauthorized users from accessing or modifying data in ways that could introduce bias or compromise model performance.

CHAPTER 5   ALIGNING AI GOVERNANCE WITH OTHER INTERNAL GOVERNANCE MODELS FOR
             TRUSTWORTHY AI: "THE CONVERGENCE OF GOVERNANCE FRAMEWORKS"

User Access Reviews and Auditing → Regularly reviewing user access privileges and auditing access logs helps identify any suspicious activity or potential unauthorized access attempts. This proactive approach ensures access rights align with current user roles and responsibilities.

Model Security

- Secure development lifecycle (SDLC) for AI models
- Input validation to prevent adversarial attacks
- Techniques to protect models from theft or unauthorized modification

Model security is essential to protect an AI model's intellectual property and safeguard the integrity and reliability of its outputs. Compromising a model used in critical decision-making could have severe consequences, ranging from financial losses to reputation damage and even harm to individuals.

Secure Development Lifecycle (SDLC) → Implementing a secure SDLC for AI models integrates security considerations throughout the entire development process, from initial design to deployment.

Sanitizing Inputs → Ensure all data fed into an AI model is rigorously validated to prevent adversarial attacks. These attacks often involve manipulating input data to fool the model into making incorrect predictions or revealing sensitive information.

Error Handling → Implement robust error-handling mechanisms to prevent unexpected inputs from system failures or data exfiltration.

Threat Modeling → Proactively identify potential attack vectors and vulnerabilities specific to AI models. This includes scenarios like adversarial attacks, model theft, or unauthorized modification.

Secure Coding Practices → **MLOps and developers must be trained** in secure coding techniques to prevent common vulnerabilities (like buffer overflows and SQL injection) that could be exploited to modify the AI model or gain access to sensitive data.

Code Reviews and Testing → To uncover potential vulnerabilities, incorporate security testing throughout the development process, including static code analysis tools and penetration testing.

Version Control and Change Management → Use secure version control systems to track changes to AI models, ensuring any modifications are authorized and can be traced back to their source for auditing and rollback in case of an incident.

Data Provenance and Lineage

- Tracking the origin and transformations of data throughout the AI lifecycle
- Ability to audit decisions (explaining model rationale)

Monitoring and Auditing

- Continuous monitoring of AI systems for anomalies and potential attacks
- Logging and auditing of data access, model changes, and decision-making

Incident Response and Disaster Recovery

- Clear incident response plan with roles and responsibilities tailored to AI risks
- Regular testing of backup and recovery procedures

Incident Response Plan. A clear and well-defined incident response plan tailored to AI-specific risks is vital. This plan outlines roles and responsibilities in case of a security breach or malfunction, ensuring a swift and coordinated response to minimize damage. The plan should include procedures for isolating the incident, containing the threat, investigating the root cause, and recovering from the event.

CHAPTER 5   ALIGNING AI GOVERNANCE WITH OTHER INTERNAL GOVERNANCE MODELS FOR TRUSTWORTHY AI: "THE CONVERGENCE OF GOVERNANCE FRAMEWORKS"

Cyberattacks targeting AI systems can steal valuable models, manipulate training data, or disrupt model outputs. Effective incident management minimizes the damage caused by such attacks, protects intellectual property, and ensures the AI system remains trustworthy. AI incidents can expose sensitive data used for training or held within AI-powered systems. Swift incident response is essential to contain breaches, prevent unauthorized data access, and comply with data privacy regulations.

It's important to note that incident management for AI goes beyond traditional IT incidents. It must address unique risks such as adversarial attacks, where inputs are manipulated to deceive the model, and incidents involving biased or unethical AI outputs. How an organization responds to AI incidents directly impacts public trust in its AI systems. Quick containment, transparent communication, and corrective action demonstrate a commitment to responsible AI use.

Key Components of AI-Specific Incident Management and Response

Tailored Incident Response Plan(s) should specifically address AI-related threats, inclusive of the following elements:

- Identifying critical AI assets and their vulnerabilities
- Roles and responsibilities for AI incident response (data scientists, security experts, etc.)
- Procedures for isolating affected AI systems and data
- Communication protocols, both internally and with external stakeholders

AI Expertise on Incident Response Teams should include data scientists, AI developers, or AI security specialists who understand the unique nuances of AI incidents. They can assist in isolating the problem, assessing potential bias implications, and identifying the root cause for remediation. Post-incident analysis must go beyond technical root causes.

Investigate how AI models might have been targeted, whether the incident affected the fairness or integrity of AI decisions, and how such incidents can be better prevented in the future.

Organizations should foster strong collaboration between security teams, AI developers, and compliance officers. This proactive approach ensures incident response plans are aligned with data governance policies and considers the legal and ethical implications of AI-related security incidents.

Successful AI incident management prioritizes speed, understanding that timely responses minimize damage to the AI system and public trust. Having AI expertise readily available to assist with isolation, analysis, and remediation is essential. Ensure incident response plans are aligned with data governance policies and consider legal and ethical ramifications. Furthermore, compliance with data privacy regulations regarding breach reporting is non-negotiable. Finally, embrace a continuous learning mindset by analyzing past incidents to proactively strengthen defenses and share threat intelligence within your organization and the broader AI community.

Backup and Recovery Procedures. Regularly backing up AI models, training data, and system configurations allows for swift recovery in case of a cyberattack or system failure. Testing backup and recovery procedures ensures they are functional and can be executed efficiently to minimize downtime and data loss. Backup and recovery procedures are paramount for AI governance and risk management as they safeguard invaluable AI models, the data they rely on, and system configurations. This minimizes system downtime and data loss in the face of adversarial AI and cyberattacks, system failures, or even accidental data deletion. I stress the importance of testing backup and recovery processes to ensure swift restoration, preserving the integrity of AI-driven operations, and mitigating risks associated with lost intellectual property or the inability to deliver expected AI outputs reliably. Backup and recovery reinforce responsible AI governance by demonstrating preparedness and prioritizing the continuity of AI systems. The core components of an effective backup and recovery strategy include the following:

- Prioritize what needs to be backed up. This includes AI models, training and testing datasets, system configurations, code repositories, and any metadata associated with the AI development process.

- Determine how often backups occur based on the rate of change for critical assets and your organization's risk tolerance. Highly dynamic models and data might necessitate more frequent backups than stable systems.

- Store backups in multiple locations, ideally using on-site and secure off-site locations (cloud-based solutions often provide geographic redundancy). This safeguards against localized disasters or single points of failure.

- Protect backups with encryption and strict access controls to prevent unauthorized access or modification in the event of a security breach.

- Maintain multiple backup versions to allow recovery to a known good state in case of data corruption or model degradation.

- Regularly test recovery procedures to ensure they function as expected and that you can restore data and systems within acceptable time frames.

- Backup strategies should include tracking the data's lineage and any transformations used for model training. This aids in auditing and understanding how models were affected at a particular point in time.

- If possible, consider backing up any explainability tools or metadata alongside the AI models to facilitate root cause analysis and remediation in case of an incident.

- Incorporate recovery procedures into your AI-specific incident response plan to ensure swift restoration of critical systems.

A well-defined backup and recovery strategy is not just a safeguard; it's a core pillar of responsible AI governance that minimizes downtime and ensures the long-term resilience of your AI initiatives.

Managing Cybersecurity Risks AI Systems Face

AI systems face a complex threat landscape, encompassing both traditional cybersecurity risks and those unique to their data-driven nature. Adversaries can attack the data supply chain, poisoning training data to induce errors or biases in the AI model. Malware and network intrusions continue to pose threats, allowing attackers to disrupt operations, exfiltrate sensitive data, or compromise the AI model itself. AI-specific attacks are particularly insidious. Adversarial examples leverage the sensitivities of AI models to deceive them into misclassifications or incorrect decisions. Model inversion and model theft raise concerns about privacy and intellectual property loss. Executives must recognize that AI systems introduce new attack surfaces. Vulnerabilities in data pipelines, the models themselves, and any associated APIs provide potential entry points. AI security demands a proactive approach, recognizing the trade-offs between security and model performance and the need to adapt defenses against evolving adversarial techniques continuously.

Due to their unique nature, AI systems face a range of cybersecurity attack vectors. Here's a breakdown of the usual suspects:

Table 5-1. AI Attack Vectors and Mitigants

Attack Vector	Attack Type	Risk Factor	Mitigating Controls
Traditional	Data Poisoning	Adversaries can subtly manipulate your training data to skew the AI model's behavior, leading to incorrect or biased outputs.	Validate/sanitize input data to prevent malicious data from entering the training set, utilizing outlier removal, normalization, and flagging suspicious entries.
	Malware	AI systems can be infected with traditional malware, such as ransomware, disrupting operations or allowing attackers to steal data or models.	Endpoint protection, vulnerability patching, input sanitation, whitelisting, access controls, monitoring, and an incident response plan to shield AI systems from traditional malware.
	Data Supply Chain	Data collection, storage, and processing vulnerabilities provide potential entry points for attackers.	Implement security measures throughout every data collection, storage, and processing stage to mitigate supply chain vulnerabilities and reduce the risk of attacks.

(*continued*)

Table 5-1. (*continued*)

Attack Vector	Attack Type	Risk Factor	Mitigating Controls
	AI Models Themselves	Adversaries can exploit model design, implementation, or deployment flaws.	Adopt testing, code reviews, secure deployment, and continuous monitoring to minimize the risk of adversaries exploiting vulnerabilities within the AI model's lifecycle.
	APIs and Interfaces	APIs that interact with AI systems may introduce vulnerabilities if not properly secured.	Implement strict authentication, authorization, input validation, and monitoring for all APIs interacting with AI systems.
	Network Intrusions	Attackers exploiting vulnerabilities can gain unauthorized access to AI systems, steal models, or tamper with the data used for training or decision-making.	Implement network security measures to safeguard AI systems from unauthorized access, data theft, and manipulation.
	Social Engineering	Using phishing or pretexting, attackers can trick users into executing malicious code targeted at AI systems.	Educate all users on social engineering threats to protect AI systems.

(*continued*)

Table 5-1. (*continued*)

Attack Vector	Attack Type	Risk Factor	Mitigating Controls
Adversarial AI	Adversarial Attacks	Attackers craft malicious inputs that appear innocuous to humans but fool AI models into making incorrect predictions or classifications.	Utilize adversarial training techniques and robust input validation to protect AI models from adversarial attacks.
	Model Inversion Attacks	Attackers with access to an AI model's outputs and knowledge of its architecture can reconstruct sensitive training data, exposing privacy risks.	Employ differential privacy and restrict model query access to combat model inversion attacks and safeguard sensitive data.
	Model Theft	Attackers can steal proprietary AI models through various means, losing intellectual property and potentially giving competitors an advantage.	Protect AI models with encryption, access controls, and watermarking techniques to deter and detect model theft.
	Evasion Attacks	Adversaries can design inputs to bypass security mechanisms intended to protect AI systems (e.g., crafting malware that evades detection by traditional antivirus software).	Implement continuous monitoring, adversarial training, and anomaly detection to identify and defend against evasion attacks on AI systems.

CHAPTER 5   ALIGNING AI GOVERNANCE WITH OTHER INTERNAL GOVERNANCE MODELS FOR TRUSTWORTHY AI: "THE CONVERGENCE OF GOVERNANCE FRAMEWORKS"

AI Model Risk Management

AI Model Risk Management

AI model risk management requires a proactive approach woven throughout the entire AI development and deployment process. Prioritize risk identification, implement technical and ethical safeguards, establish robust governance, and foster ongoing communication and adaptation. Investing in principled AI risk management protects against potential harm, promotes trust, and lays the foundation for long-term AI success.

So, what is the difference between AI Model Risk Management and Cyber or Enterprise Risk Management?

Traditional risk management centers on protecting IT infrastructure from data breaches, safeguarding financial assets, and upholding operational continuity – think firewalls, access controls, and disaster recovery plans. In contrast, AI Model Risk Management dives deeper, tackling unique risks inherent in AI systems. The focus shifts to mitigating bias, ensuring fairness and explainability, maintaining data quality, and preventing unintended consequences. To address these complex and potentially less quantifiable risks, technical safeguards must be paired with ethical oversight and governance practices designed to instill trust in AI.

Organizations need a risk management strategy beyond conventional frameworks to responsibly tap into AI's transformative power. This strategy must address the distinct technical, ethical, and reputational risks of AI systems. A proactive approach that identifies vulnerabilities, prioritizes technical and ethical safeguards, establishes transparent governance, and embraces ongoing learning is key to fostering trustworthy AI and avoiding potential harm. Ensure documented processes and procedures are in place to determine the needed level of risk management activities based on the organization's risk tolerance.

Successful AI model risk management hinges on open communication and a commitment to continuous learning. Fostering transparency throughout the organization, from developers to executives, builds a shared understanding of the risks AI poses and the strategies to address them. This includes open dialogue and a culture where reporting potential issues is encouraged.

Because AI threats and regulations evolve rapidly, your risk management framework must be equally adaptable. Proactively update it in response to industry shifts, emerging threats, and lessons learned within your organization. Collaboration is crucial – engage with industry groups and share threat intelligence to stay ahead of potential dangers. A holistic approach is essential, addressing technical, ethical, reputational, and regulatory risks. Ensure your strategy grows alongside your AI initiatives and always aligns with trusted ethical AI frameworks to ensure responsible and trustworthy AI use.

Successful AI risk management demands more than just technical fixes. Proactively address the ethical implications, potential reputational damage, and evolving regulatory requirements that AI models introduce. Keep pace with scaling AI adoption by designing a risk management strategy that grows and adapts alongside increasing model complexity. Most importantly, ground your approach in an established ethical AI framework. This provides a strong foundation for making responsible and trustworthy choices, shaping how to mitigate risks and, ultimately, how to deploy AI technologies.

To implement intelligent AI model risk management, begin by understanding your organization's AI landscape. This involves creating an inventory of all AI models, noting their purpose, criticality, and the sensitivity of the data they utilize. Consider how each model is deployed and the potential consequences of errors or misuse, develop a clear risk framework that categorizes the key risks (technical, ethical, reputational, and regulatory), and establish a prioritization system based on your risk tolerance to guide your mitigation efforts. Specific risk assessment methods include:

Identifying and Assessing AI Model Risks

Proactive Threat Modeling → Identify potential risks specific to your AI models, spanning technical flaws, data biases, adversarial attacks, unintended consequences, and ethical concerns.

Bias Assessment → Use fairness metrics and bias detection tools to analyze datasets and evaluate model outputs for potential discrimination across protected groups.

Scenario Analysis → Consider "what if" scenarios to understand the impact of data shifts, adversarial attacks, or ethical dilemmas.

Risk Prioritization Framework → Develop a system to categorize risks based on their likelihood and potential impact. Focus on high-priority risks, ensuring mitigation efforts align with your organization's risk appetite.

Mitigation and Control

Secure Development Lifecycle (SDLC) Risks → Embed security and risk considerations throughout the AI development process. Enforce secure coding, rigorous testing, and privacy by design principles. More will be discussed in Chapter 21.

Data Governance Risks → As we have discussed in the data governance section of this book, organizations should implement strong data governance practices encompassing data quality checks, bias detection, data provenance tracking, and robust access controls.

Model Validation and Monitoring Risks → Establish thorough validation before deployment and continuous monitoring to detect model drift, performance degradation, or unexpected biases.

Explainability Risks → Where applicable, leverage methods like LIME and SHAP to enhance understanding of model decision-making and facilitate troubleshooting and auditing.

Incident Risks → Develop an AI-specific incident response plan addressing data breaches, model tampering, or scenarios of biased/unethical AI outputs.

CHAPTER 5　ALIGNING AI GOVERNANCE WITH OTHER INTERNAL GOVERNANCE MODELS FOR TRUSTWORTHY AI: "THE CONVERGENCE OF GOVERNANCE FRAMEWORKS"

AI Risk Governance and Oversight

AI Risk Committee → Establish a cross-functional committee (data scientists, compliance, IT, and business stakeholders) to oversee AI model risk management.

Clear Roles and Responsibilities → Define ownership and accountability for identifying, assessing, mitigating, and reporting AI risks.

Audits and Reviews → Implement regular independent audits and reviews of AI models (covering technical, ethical, and compliance aspects).

I provide clients with a few strategic recommendations that may resonate with their planned or conceptual approaches.

AI Model Risk Management High-Level Recommendations

- **Establish mechanisms to regularly review the efficacy of your AI governance and risk management processes**, i.e., *perform regular assessments of your AI governance and risk management processes based on model risk and business criticality. (Risk-based Approach)*

- **Establish a process for ranking and tiering models based on risk level and business criticality.** For example, *models that make business decisions based on personal or sensitive data or models that make decisions based on demographics should be ranked by risk and tier (Tiered 1-5 (1 being highest) and High Risk 5-1 (5 being highest). Key risk indicators should be identified, used, and incorporated into your enterprise risk management structure.*

- **Verify that formal AI risk management policies align with existing legal and cybersecurity standards,** industry best practices, and norms. Specifically, AI risk management policies should be established to align with AI and model control procedures and verified that currently deployed and third-party AI models conform.

- **Establish policies to define mechanisms for measuring or understanding the AI model's potential impacts.** For example, a procedure should be established to perform regular impact assessments at key stages in the AI model development lifecycle connected to system impacts and the frequency of system updates.

- **Establish policies and processes that define impact/risk assessment scales for measuring the potential impact of the AI system.** *Red-amber-green (RAG) scales may be qualitative or entail simulations or econometric approaches.*

- **Establish policies for assigning an overall risk measurement approach for an AI model or its important components**, *i.e., via multiplication or combination of a mapped risk's impact and likelihood (risk ≈ impact x likelihood).*

- *Identify AI model (RACI) actors responsible for evaluating the efficacy of risk management processes and approaches and adjusting courses based on the results.*

- **Establish policies and processes regarding internal and public disclosure (where applicable) of the use of AI and risk management material,** *i.e., impact assessments, audits, model documentation, and validation and testing results.*

- **Develop and document procedures** for the ongoing monitoring and periodic review of the risk management process and ensure its outcomes are planned, organizational roles and responsibilities are clearly defined, including determining the frequency of systematic reviews, i.e., *establish policies to allocate appropriate resources and capacity for assessing impacts of AI models on individuals, communities and society and policies and procedures for monitoring and addressing AI model performance and trustworthiness, including bias and security problems, across the lifecycle of the model.*

- **Implement or align policies to define organizational functions and personnel responsible for AI model incident response activities**, *i.e., policies for AI model incident response or confirm that existing incident response policies apply to AI models.*

- **Develop and implement procedures and standards for AI model development within various business units.** For example, isolated and centralized cybersecurity and model development controls can be developed to ensure that the tools and *procedures within the model lifecycle meet the requirements and objectives of trusted AI.*

- **Establish a process to regularly review documentation, controls, and policies** that, among others, address information related to

  - Expected and potential risks and impacts
  - AI actor's contact information

- Business justification and use case
- Scope and usage of AI Models
- Assumptions and limitations
- Description and characterization of training data
- Algorithmic methodology
- Evaluated alternative approaches
- Description of output data
- Testing and validation results (including explanatory visualizations and information)
- Down- and upstream dependencies
- Plans for deployment, monitoring, and change management
- Stakeholder engagement plans

- **Implement processes and procedures for securely decommissioning and phasing out of AI models safely** and in a manner that does not increase risks or decrease the organization's trustworthiness, *i.e., policies that delineate where and for how long decommissioned systems, models, and related artifacts are stored and policies that address ancillary data or artifacts that must be preserved for fulsome understanding or execution of the decommissioned AI model, e.g., predictions, explanations.*

- **Implement accountability structures so that the appropriate teams and individuals are empowered, responsible, and trained to map, measure, and manage AI risks.** *Document roles and responsibilities (RACI) and lines of communication related to mapping, measuring, and managing AI model risks. Ensure they are documented and clear to individuals and teams throughout the organization.*

- **Take a risk-assessment-by-design approach and integrate this approach early in the AI development lifecycle**, i.e., *develop a model lifecycle development process that includes process risk and controls within model development, validation and approval, model use case deployment, and ongoing monitoring.*

Another excellent question from my clients is, "How do I incorporate Key Risk Indicators (KRIs) into my AI model risk strategy?" My documented responses are in Table 5-2.

Table 5-2. KRI Management

What to do with KRIs	How to with KRIs
Tailor KRIs to *Your* AI Landscape Don't Just Copy	Avoid generic KRIs. Reflect on your organization's specific AI use cases, risk appetite, and the regulatory environment you operate within.
Map KRIs to Risk Categories	Link your KRIs directly to your established risk taxonomy (technical, ethical, reputational, regulatory). This ensures your monitoring has a clear purpose.

(*continued*)

CHAPTER 5   ALIGNING AI GOVERNANCE WITH OTHER INTERNAL GOVERNANCE MODELS FOR TRUSTWORTHY AI: "THE CONVERGENCE OF GOVERNANCE FRAMEWORKS"

Table 5-2. (*continued*)

What to do with KRIs	How to with KRIs
Balance Leading and Lagging Indicators/ Leading Indicators	Focus on proactive metrics that signal potential issues before they escalate, such as rising error rates, increasing data drift, or an uptick in customer complaints hinting at fairness concerns.
Lagging Indicators	Track historical incidents and outcomes (e.g., number of security breaches, confirmed bias events). While they reflect past events, they inform your risk assessments and reveal where you might need stronger controls.
Think Beyond Traditional Metrics with Data Quality KRIs	Measure data completeness, consistency, and potential for bias across sensitive variables.
Model Explainability KRIs	Track how easily your models' decisions can be understood and the level of insight your explainability tools provide.
Ethical Alignment KRIs	Develop metrics that reflect your organization's ethical principles, potentially focused on fairness disparities or societal impact assessments.
Establish Clear Thresholds and Reporting by Defining What's "Abnormal" vs. What's "Normal"	Set meaningful thresholds for each KRI that trigger alerts and potentially further investigations.
Contextualize KRIs	Consider relevant trends or external factors that might influence KRI interpretation.

(*continued*)

Table 5-2. (*continued*)

What to do with KRIs	How to with KRIs
Create Actionable Reporting	Design dashboards or reports that communicate risk levels to stakeholders, facilitating timely decision-making.
Embrace Continuous Improvement Start Simple, Then Iterate	Begin with a core set of impactful KRIs and expand as your AI risk management matures.
Feedback Loop	Use insights from KRI monitoring to refine risk assessments, prioritize mitigation efforts, and adjust the KRIs.
Embrace Collaboration	Involve data scientists, security experts, and compliance specialists in defining, monitoring, and interpreting your KRIs.

Integrating key risk indicators (KRIs) into your AI risk management program provides a powerful tool for avoiding potential problems. Remember, success lies in tailoring your KRIs to the unique risks your AI models face and your organization's specific priorities – balance leading indicators for proactive risk detection with lagging indicators for historical insight. Do not be afraid to define innovative KRIs that track data quality, explainability, and alignment with your ethical AI principles. Finally, view your KRI framework as a dynamic tool – establish clear reporting mechanisms, use insights from KRI monitoring to continuously improve your risk management program, and foster collaboration across teams for maximum impact.

Here is a sample list of Key Risk Indicators (KRIs) for AI model risk management, broken down into categories:

Technical Performance

- Accuracy degradation → The percentage change in accuracy metrics (e.g., precision, recall, F1-score) over time.

- Rate of unexpected outputs → The frequency of outputs that deviate significantly from expected patterns or results.

- Concept drift: The degree of change in the relationship between input features and output predictions.

- Explainability metrics → Measures related to how well the model's decision-making process can be understood (these will vary depending on the methods used).

Bias and Fairness

- Disparate impact → The outcomes or predicted scores across demographic groups or protected categories differ.

- Statistical parity → The similarity of outcomes across separate groups.

- Counterfactual fairness: The degree to which changing a sensitive attribute would alter a prediction.

Regulatory and Ethical Considerations

- Privacy incidents → The number of data breaches, unauthorized access, or potential misuse of sensitive information.

- Compliance violations → Non-compliance with relevant regulations (GDPR, CCPA, or industry-specific regulations).

- Incidents of unfair or discriminatory outputs → Documented cases where the AI model produces biased results with potential harm.

- Stakeholder feedback → The number of negative sentiments or concerns raised by customers, employees, or the broader public regarding the AI system.

Security and Robustness

- Adversarial attack success rate → The percentage of adversarial inputs that successfully fool the model.

- Mean-Time-to-Detection and Remediation (MTTDR) → The time taken to identify and address security vulnerabilities or other disruptions.

Operational

- Downtime or availability issues → The frequency and duration of unplanned outages or reduced AI system performance.

- Deployment failure rate → The percentage of AI models that fail during deployment or integration.

Important Notes to Consider

Tailor to your context → Choose KRIs most relevant to your specific AI use cases, data sensitivity, and regulatory landscape. *Establish thresholds:* Define thresholds for each KRI to trigger alerts or a more in-depth investigation. *Track trends:* Monitor KRIs over time to identify potential issues early and take corrective action.

CHAPTER 5   ALIGNING AI GOVERNANCE WITH OTHER INTERNAL GOVERNANCE MODELS FOR TRUSTWORTHY AI: "THE CONVERGENCE OF GOVERNANCE FRAMEWORKS"

AI Model Risk Process Recommendations

Artificial Intelligence offers tremendous business competitive advantages and opportunities to reduce quantitative and qualitative resource requirements. However, AI can be particularly susceptible to a wide range of AI-related risks through all phases of the AI lifecycle. For example, AI-based systems may introduce or reinforce a risk of perpetuating inequity and historical bias, and enforceable regulations to protect the public by ensuring equitable, ethical, and trustworthy/transparent AI may be critical and inevitable. I recommend a multi-pronged approach for effective AI governance and model risk management that addresses all the details about your models across their lifecycle.

Preliminary Model Risk Identification and Assessment

Post the model proposal stage, I recommend (as part of the pre-validation process) a high-level preliminary risk identification stage per the following:

- The inherent risk identification can be a basic risk categorization, i.e., Tier 1-5, HML, critical/non-critical, etc.

Figure 5-4 below illustrates the recommended inherent risk strategy in alignment with AI Governance.

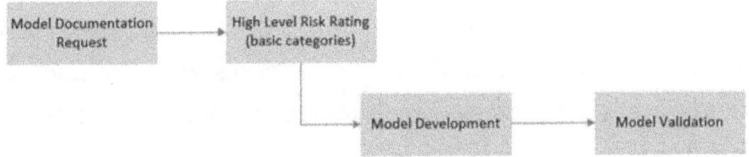

Figure 5-4. *Inherent Risk Identification Process*

Developing a risk library is recommended where all model risks are identified, logged, and managed. Later stages can include additional categorizations, risk factors, calculations, and methods, as illustrated in Figure 5-5.

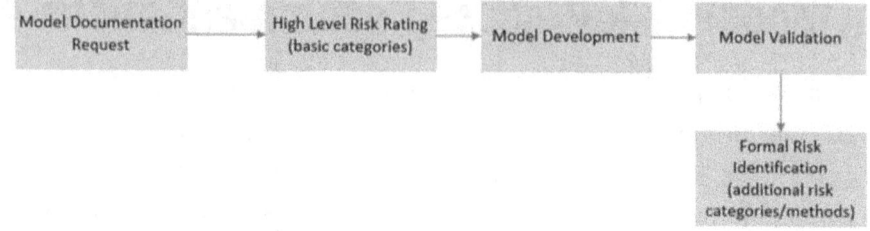

Figure 5-5. Additional Risk Categorization

AI Governance Metric Definition and Reporting

Why are AI governance metrics so important?

AI governance metrics are essential because they hold AI systems accountable to ethical standards and organizational core values. They allow you to proactively mitigate risks, track performance, and ensure alignment with ethical principles. To measure AI governance effectively, start by outlining the most important dimensions of your organization (fairness, transparency, security, etc.). Then, pinpoint specific metrics under each dimension, establish systems for data collection, and determine benchmarks for healthy performance. Transparently communicate these metrics through reports and visualizations.

Remember, AI governance is a dynamic process. Engage with diverse viewpoints, tailor metrics to your specific use cases, and continuously review and adapt your strategy to maintain the responsible use of AI within your organization.

There are several methods for collecting and reporting on AI-based metrics. However, I provide the following table of recommendations to assist you with the foundational principles of defining and reporting on AI-based metrics. I also provide a priority level for each one. See Table 5-3.

Table 5-3. *AI-based Metrics*

Metric Category	Rationale	Priority
Accountability	Metrics provide tangible evidence of whether AI systems uphold company values and align with regulatory standards. They create transparency and build trust.	*High*
Risk Mitigation	By tracking metrics like bias, system failures, or unintended consequences, organizations can proactively identify and manage potential risks before they become significant problems.	*High*
Performance Tracking	Metrics allow for evaluating the effectiveness of AI systems. Are they delivering on their intended purpose? Are they cost-effectively producing value? Metrics enable informed decision-making about optimization and improvement.	*Medium*
Ethical Considerations	Metrics tailored to fairness, explainability, and non-discrimination promote the development and use of AI that aligns with ethical principles.	*Medium*

AI governance metrics are essential to the organization. They provide quantifiable measures to ensure the ethical, responsible, safe, and transparent deployment of AI systems, mitigating risks and protecting the organization's reputation.

Strategy to Measure AI Governance Metrics

Identify Key Dimensions. Begin by mapping out the most critical governance areas for your organization. These may include

- Fairness and Bias
- Explainability and Transparency
- Security and Robustness
- Accountability and Responsibility
- Value Alignment (with company mission and goals)

Define Specific Metrics → Under each dimension, determine the metrics most relevant to your AI use case, for example:

Fairness → Disparity rates across demographic groups, statistical tests for bias.

Explainability → Decision tracing tools, importance weights for input features.

Security → Vulnerability scans, offensive security testing/purple-team testing, incident tracking.

Accountability → Audit trails and clear ownership structures.

Value Alignment → Cost-benefit analysis, ROI, impact on KPIs.

Data Collection and Tracking → Establish processes to collect necessary data. This might involve log files, user feedback mechanisms, or direct monitoring tools. Build a system for regularly tracking and storing these metrics.

Benchmarks and Thresholds → Determine what constitutes acceptable performance for each metric and set thresholds to trigger reviews or corrective actions.

Reporting and Visualization → Create dashboards or reports that communicate metric performance to stakeholders. Use visualizations to make the metrics easily understandable.

Continuous Review and Adaptation → AI governance is an ongoing process. Regularly evaluate your metrics and strategies and adapt them as technologies and best practices evolve.

Important Considerations

Involve Diverse Stakeholders → When designing your governance metrics, engage with technical teams, business leaders, ethicists, and potentially even end users.

Context Matters → Tailor your metrics to the specific AI systems and risks relevant to your organization.

Strive for Balance – While thorough governance is paramount, avoid overly restrictive metrics that stifle innovation.

AI Governance Key Performance Indicators (KPIs)

In this book, I've repeatedly defined AI governance as the framework of policies, processes, and controls that guide the development, validation, deployment, and use of AI models within an organization, with an assurance that AI is implemented ethically, responsibly, and in a way that aligns with the organization's goals. On the other hand, AI governance and model risk management KPIs (Key Performance Indicators) are measurable metrics that track the effectiveness of your AI governance and model risk program. KPIs can be categorized into different areas like fairness, security, and performance. By monitoring these KPIs, your organization can identify and address potential risks associated with AI models, such as data drift, model hallucinations, security vulnerabilities, and performance degradation.

My recommendations on KPI strategies and best practices.
Establish a Clear Framework

CHAPTER 5   ALIGNING AI GOVERNANCE WITH OTHER INTERNAL GOVERNANCE MODELS FOR TRUSTWORTHY AI: "THE CONVERGENCE OF GOVERNANCE FRAMEWORKS"

Assemble a cross-functional governance team with representatives from various departments (IT, business, legal, compliance, risk) to set standards and oversee AI initiatives. Define and create policies and procedures addressing data governance and handling, model development, model validation, model risk management, bias prevention, explainability, security, and change management.

Prioritize Transparency and Accountability:

Maintain detailed (documented) records on data sources, model development processes, decisions, and performance monitoring for accountability and auditing. Ensure models, especially high-risk ones, can be explained to understand how decisions are made.

Focus on Ethical Considerations

[I know you are probably tired of seeing ethical considerations in this book, but trust me, there are compliance-driven reasons for me stressing it].

Proactively evaluate models for biases and implement mitigation strategies to reduce discriminatory outcomes. Before deploying AI systems, assess the impact on society and address potential issues. Review your governance framework, policies, and KPIs regularly to align with evolving regulations and technology and train employees across the organization on AI ethical principles, governance, and risk management.

Implement Comprehensive Model Risk Management Procedures

Model validation and testing. Use rigorous validation processes before deployment and conduct ongoing testing to ensure performance and robustness. Monitor key model performance metrics in production and establish methods for addressing issues.

CHAPTER 5   ALIGNING AI GOVERNANCE WITH OTHER INTERNAL GOVERNANCE MODELS FOR
            TRUSTWORTHY AI: "THE CONVERGENCE OF GOVERNANCE FRAMEWORKS"

Key KPIs to Consider:

- **Fairness** → Metrics measuring disparate impact across protected groups

- **Explainability** → Measures of how well model decisions can be explained

- **Accuracy and Performance** → Traditional metrics like precision, recall, and F1-score

- **Robustness and Stability** → Metrics assessing a model's resilience to data changes or adversarial attacks

- **Data Quality** → Metrics on data completeness, consistency, and bias

- **Documentation and Auditability** → Completeness of records for model lineage and change tracking.

In summary, effective AI governance and risk management require a collaborative effort. By following these best practices and carefully selecting KPIs, your organization can establish a sturdy foundation to build trustworthy, responsible AI systems.

Measure the appropriateness of AI metrics and effectiveness of existing controls through regular assessments and including reports of errors and impacts on affected communities or individuals, *i.e., assess the external validity of all measurements (e.g., the degree to which measurements taken in one context can generalize to other contexts) and scope for sharing metrics and related information with compliance authorities, stakeholders, and impacted communities.*

CHAPTER 5 ALIGNING AI GOVERNANCE WITH OTHER INTERNAL GOVERNANCE MODELS FOR TRUSTWORTHY AI: "THE CONVERGENCE OF GOVERNANCE FRAMEWORKS"

Summary and Thoughts

An effective strategy for aligning AI governance with other organizational governance models lies in establishing a common foundation of core values (transparency, accountability, fairness) and adopting an integrated risk management framework. This involves mapping existing policies across data governance, ethical principles, cybersecurity, and privacy to identify gaps or redundancies specifically related to AI use cases. It requires creating cross-functional working groups to bridge communication between technical and governance teams, ensuring consistent standards and a shared understanding of risks. Finally, regular reviews and updates should be implemented to this integrated governance framework, ensuring it remains agile and responsive to both the evolving nature of AI and the changing regulatory landscape.

Quiz

1. What is a key characteristic of trustworthy AI integrated into organizational policies, processes, and procedures?

    A) High performance

    B) Rapid deployment

    C) Ethical design

    D) Minimal oversight

CHAPTER 5 ALIGNING AI GOVERNANCE WITH OTHER INTERNAL GOVERNANCE MODELS FOR TRUSTWORTHY AI: "THE CONVERGENCE OF GOVERNANCE FRAMEWORKS"

2. Which governance model involves fairness, non-discrimination, transparency, accountability, privacy, security, and social benefit?

   A) AI Security Governance

   B) Data Governance

   C) AI Ethics

   D) Model Risk Management

3. What is essential for building trust and mitigating risks in AI systems?

   A) Rapid development

   B) AI Security Governance

   C) Minimal documentation

   D) Limited stakeholder engagement

4. Which framework proactively identifies potential harms like discrimination, security vulnerabilities, or operational failures?

   A) Explainability framework

   B) Risk assessment framework

   C) Model validation framework

   D) Data quality framework

5. True or False: AI Governance ensures that AI aligns with organizational goals, ethical values, and legal regulations.

6. True or False: Data governance for AI does not involve data lineage.

CHAPTER 5 ALIGNING AI GOVERNANCE WITH OTHER INTERNAL GOVERNANCE MODELS FOR TRUSTWORTHY AI: "THE CONVERGENCE OF GOVERNANCE FRAMEWORKS"

7. True or False: AI Governance is only about implementing technology and does not involve processes or people.

8. AI Governance and Model Risk Management bring accountability and _____ to AI/ML models throughout their lifecycle.

9. AI security governance and risk management are essential for building trust and mitigating _____ in AI systems.

10. Data governance for AI involves data quality, security, privacy, bias mitigation, and data _____.

CHAPTER 6

Designing Your AI Governance Framework

Frame the Goals of Building a Successful AI Governance Framework

Imagine AI as a powerful new energy source. In its early days, pioneers harnessed this energy with amazing results. But they also faced unexpected explosions, malfunctions, and unintended consequences. Now, organizations that want to leverage AI responsibly find themselves in a similar landscape. Early adopters have learned valuable lessons, but much remains uncharted territory.

Why AI Governance Matters

Think of AI governance as building a system of power lines, safety protocols, and regulations for this new energy source. It aims to

Channel Innovation Responsibly → Instead of AI projects running wild, a framework helps direct them toward solving real problems in ways that align with ethical principles.

Minimize Risk → Just like electrical failures can be costly, AI misfires can have legal, financial, and reputational damages. Governance is like an insurance policy for your organization.

Build Trust as a Competitive Edge → Customers are warier of "black box" AI. A well-communicated governance framework reassures users, partners, and regulators, increasing trust in your brand.

Navigate a Complex Future → AI regulation is not standardized. Your governance becomes your compass, helping you adapt as the legal landscape around AI evolves.

What a Successful Framework Looks Like

Picture it not as a cage for AI but more like the bones of a healthy skyscraper:

Strong Foundation → Clear ethical principles and a commitment from leadership lay the groundwork.

Flexible Structure → Policies and processes allow various AI projects to flourish within guardrails.

Active Monitoring → Systems for detecting bias, model drift, and security threats are crucial – not just building the framework but ensuring it works in real time.

Built to Evolve → Successful frameworks are reviewed and updated, learning from experience and new regulations.

Key Point!

AI governance is an investment in your organization's long-term health. It enables you to scale your AI projects confidently, secure a reputation for responsibility, and ultimately deliver more positive, impactful outcomes to society.

CHAPTER 6   DESIGNING YOUR AI GOVERNANCE FRAMEWORK

A Practical Guide for Building Your AI Governance Framework

How do I start building my framework?

Here's a practical guide and three-phase process on how to start building your AI governance framework, focusing on actionable first steps:

Phase 1: Foundational Steps

- **Secure Executive Buy-In** → Explain the benefits of AI governance (risk mitigation, competitive advantage, ethical alignment) to secure leadership support and resources. A senior-level champion is crucial.

- **Assemble a Cross-Functional Team** → Form a starting governance committee that includes representatives from data science, IT, legal, risk, and business units using AI, as well as potentially an ethical advisor.

- **Initial Landscape Assessment** → Get a baseline understanding of

    - Current AI projects and use cases across the organization.

    - Existing data management practices.

    - Relevant regulations that already apply to your organization.

- **Core Design Principles Tailored to Your Reality** → Your AI governance needs to reflect your organization's size, industry, risk tolerance, and the current maturity of your AI projects.

CHAPTER 6   DESIGNING YOUR AI GOVERNANCE FRAMEWORK

- **Actionable, Not Academic** → Focus on clear guidelines and processes that teams will use, not just abstract principles.

- **Collaborative** → Treat governance as a company-wide effort, soliciting input from data scientists, business stakeholders, legal, and those potentially impacted by AI decisions.

- **Built to Evolve** → Technology and regulations change rapidly. Embed regular review cycles into your framework's DNA.

Phase 2: Developing Core Components

- **Sponsorship** → Secure backing from a senior executive who champions the value of responsible AI.

- **Governance Team** → Assemble a cross-departmental group to draft the initial framework and ensure diverse perspectives.

- **Baseline Assessment** → Map current AI projects, data practices, and relevant regulations (even general ones like GDPR).

- **Articulate Ethical Principles** → Draft a set of core principles (fairness, transparency, privacy, non-maleficence) that will guide your organization's use of AI.

- **Outline Key Policies** → Start with the most pressing areas, prioritizing:

- **Data Governance** – Covering data collection, storage, access, and quality.

- **Model Development and Validation** → Outlining standards, testing, and documentation.

- **Bias Mitigation** → How to proactively assess and address potential biases.

- **Risk Assessment Methodology** → Create a framework to identify, quantify, and prioritize AI-related risks (reputational, operational, legal) across the project lifecycle.

Phase 3: Defining Your Pillars

- **Ethical Foundations** → Distill your core AI principles (fairness, accountability, etc.). These guide all other framework elements.

- **Key Policies** → Start with high-priority areas:
    - Data governance (handling, permissions, bias checks)
    - Model development standards (documentation, testing, explainability)
    - Risk assessment (framework for identifying and quantifying potential harms)
    - Incident response (what to do if models fail or cause discriminatory outcomes)

Phase 4: Implementation and Iteration

- *Pilot with a Project* → Select a representative AI project to apply the draft framework, test your policies, and refine them based on real-world experience.

- *Training and Communication* → Educate relevant employees on the framework, their roles, and the importance of responsible AI.

- *Start Simple, Then Scale* → Begin with a core set of policies and expand over time. Remember, governance is about continual improvement.

- *Tools and Templates* → Do not just dump policies on teams. Provide templates, checklists, and even software tools if the budget allows.

- *Awareness and Training* → Educate relevant teams on the framework, why it matters, and their specific responsibilities.

Tips for Success

- *Involve Stakeholders* → Get feedback from employees affected by the framework to enhance practicality and acceptance.

- *Do not Reinvent the Wheel* → Leverage established resources and AI governance templates to get started (many are available online).

- *Seek External Input* → Consider a workshop with an AI governance expert for initial guidance on your needs if the budget allows.

Remember This is an ongoing journey! Start with a solid foundation, then continually adapt your framework as your organization's AI capabilities mature and regulations evolve.

CHAPTER 6   DESIGNING YOUR AI GOVERNANCE FRAMEWORK

Living Framework, not a Static Document

Iterative Refinement → Gather feedback from stakeholders, track how well the framework is followed, and update it regularly. *Monitor the Big Picture:* Proactively track changes in the regulatory landscape. Do not wait for laws to force you to adapt. *Communication Loop:* Transparently share successes and lessons learned with employees. This builds an organization-wide culture of responsible AI.

AI Governance Framework Model and Design Concept

Before jumping into a framework model, I offer pre-development best practices for your use and adoption.

"Right-Size," Don't Overwhelm → A complex framework can initially be discouraging. Start expanding as your AI capabilities grow. Consider an internal or external workshop or consultation with an AI ethics expert for tailored advice, especially in sensitive sectors. While meeting regulations is crucial, think beyond compliance. True ethical and transparent AI is about proactive care and minimizing the risk of harm to individuals and society. See the illustration below for a sample AI governance framework you can follow. *This model is proven and has been implemented in several of our engagements.*

CHAPTER 6 DESIGNING YOUR AI GOVERNANCE FRAMEWORK

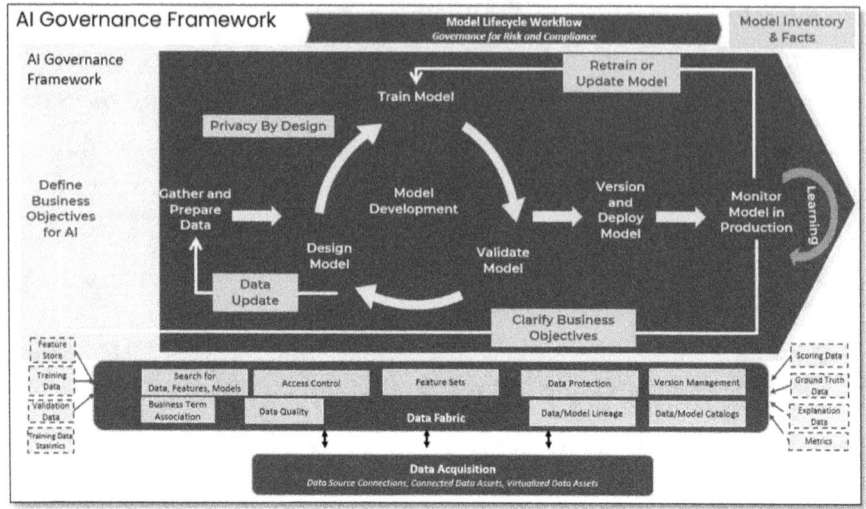

Figure 6-1. *AI Governance Framework Reference Model*

Summary and Thoughts

Designing a practical AI governance framework starts with securing strong leadership support and assembling a diverse team representing different perspectives throughout your organization. Baseline assessments of current AI use, data practices, and relevant regulations are essential. The framework itself centers on defining your ethical principles (think fairness, transparency) and crafting core policies that prioritize data governance, model development, risk assessment, and how you will handle incidents. Remember, this should not be a one-and-done process – pilot projects, tools, training, and ongoing communication will bring your framework to life.

Regularly iterate and refine your framework based on feedback and real-world experiences. Stay vigilant about the evolving regulatory landscape and foster a company-wide culture of responsible AI. Most importantly, do not just think about compliance with laws – focus on proactively building AI systems that minimize harm and genuinely deliver value to society.

CHAPTER 6   DESIGNING YOUR AI GOVERNANCE FRAMEWORK

Quiz

1. What is the first step in framing the goals of a successful AI governance framework?

    A) Implementing AI ethics guidelines

    B) Establishing a dedicated AI team

    C) Conducting a baseline assessment of current AI use and data practices

    D) Starting with pilot projects

2. Which of the following is crucial for crafting core policies in an AI governance framework?

    A) Focusing solely on data governance

    B) Ignoring stakeholder feedback

    C) Defining ethical principles such as fairness and transparency

    D) Prioritizing rapid deployment over safety

3. What is essential for bringing an AI governance framework to life?

    A) One-time training session

    B) Ongoing communication, tools, and training

    C) Restricting access to AI technologies

    D) Minimizing stakeholder involvement

CHAPTER 6   DESIGNING YOUR AI GOVERNANCE FRAMEWORK

4. Which of the following best describes the approach to refining an AI governance framework?

    A) Implementing it once and leaving it unchanged

    B) Regularly iterating and refining based on feedback and real-world experiences

    C) Avoiding regulatory updates

    D) Excluding cross-functional collaboration

5. True or False: An effective AI governance framework should only focus on compliance with current regulations.

6. True or False: Defining ethical principles is a central component of a successful AI governance framework.

7. True or False: A well-designed AI governance framework does not require baseline assessments of current AI use and data practices.

8. Designing a practical AI governance framework starts with securing strong _____ support.

9. The framework itself centers on defining your ethical principles and crafting core policies that prioritize _____, model development, risk assessment, and incident handling.

10. Most importantly, focus on proactively building AI systems that minimize harm and genuinely deliver _____ to society.

CHAPTER 7

AI Governance and Oversight Model

I decided to develop an entire chapter on the subject because AI governance and oversight are crucial for organizations that rely on decision-based AI models. AI governance and oversight are not just about compliance; they are about proactively building a responsible and sustainable AI-driven future for the organization. There is no one-size-fits-all AI governance structure. Organizations must identify roles and responsibilities at strategic, tactical, and operational levels, establish an AI governance council, and identify all groups supporting AI initiatives. Maturity, the size of the organization, and enterprise governance structure will influence AI governance structure. Here is a sample illustration of an AI governance oversight model:

CHAPTER 7   AI GOVERNANCE AND OVERSIGHT MODEL

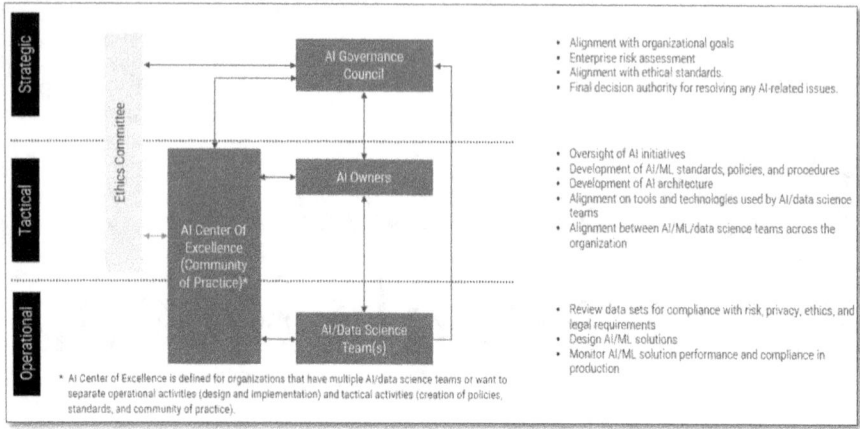

Figure 7-1.   AI Governance Oversight Model

A client recently asked me if I should develop an AI Centre of Excellence, and I gave the following recommendations and responses.

> "Creating an AI Governance Center of Excellence (CoE) can be highly valuable for your organization, particularly if AI is increasingly integrating into your business operations."

AI Governance Centre of Excellence

Why Create an AI Governance CoE?

Centralized Expertise → A CoE brings together specialists from various disciplines (data science, IT, legal, risk, ethics) to provide a dedicated focus on AI governance best practices.

Standardization and Consistency → The CoE sets organization-wide standards for AI development, data management, risk assessment, and ethical considerations, ensuring consistency across projects.

Proactive Risk Management → It identifies potential risks early in the AI lifecycle, mitigating legal, ethical, and reputational hazards.

Operational Efficiency → A CoE streamlines processes, provides tools and templates, reduces development time, and prevents costly rework caused by non-compliance.

Driving Innovation → The CoE can foster a culture of responsible AI and enable your teams to explore new use cases confidently while minimizing potential negative impacts.

The Value of an AI Governance CoE

Protection → Reduces the likelihood of legal fines, reputational damage, and public backlash due to poorly governed AI systems.

Stakeholder Trust → Builds trust with customers, partners, and regulators, who are reassured by your commitment to ethical and responsible AI.

Competitive Advantage → Responsible AI becomes a differentiator, attracting top talent and fostering stronger business relationships.

Improved Decision-Making → Access to reliable, transparent, well-governed AI models empowers better business decisions.

Long-Term Sustainability → Establishes a framework for ethical AI as its use expands within your organization, enabling sustainable growth.

...and here comes a conjunction.

BUT – Before making that decision, consider:

- **Organizational Maturity** → If you start with AI, a lightweight governance structure might suffice initially. A CoE is more beneficial as AI adoption increases and your organization's AI proficiency matures.

- **Available Resources** → Establishing a CoE requires dedicated resources and expertise. Assess if you can commit to this.

CHAPTER 7   AI GOVERNANCE AND OVERSIGHT MODEL

AI Governance Centre of Excellence Strategies

An effective AI Governance CoE strategy establishes a sharp vision and a vigorous framework of policies and processes. It provides the tools necessary to ensure responsible AI development across the organization. It focuses on risk management, compliance tracking, education, and continuous improvement. The strategy should clearly define roles and responsibilities within a cross-functional team. It should also prioritize tools for model development, testing, explainability, bias detection, and performance monitoring. Most importantly, the CoE strategy must be adaptable to the evolving regulatory landscape and foster a company-wide culture of responsible AI use.

The Core Components of an AI CoE Strategy

Vision and Mission → Clearly articulate the purpose of the CoE, aligning it with the organization's overall AI strategy and business goals. Define the scope of the CoE's authority (advisory vs. decision-making)

Governance Framework → Policies and Standards: Develop comprehensive AI principles and policies addressing data handling, model development, ethics (fairness, explainability), risk assessment, security, and compliance.

Processes → Define workflows for model development, validation, deployment, monitoring, and incident response.

Roles and Responsibilities → Establish a cross-functional team within the CoE with representation from IT, business, legal, compliance, risk management, and data science.

Governance Framework

Policies and Standards → Develop comprehensive AI principles and policies addressing data handling, model development, ethics (fairness, explainability), risk assessment, security, and compliance.

Processes → Define workflows for model development, validation, deployment, monitoring, and incident response.

Roles and Responsibilities → Establish a cross-functional team within the CoE with representation from IT, business, legal, compliance, risk management, and data science.

Tools and Infrastructure

Model Development and Testing Platforms → Provide standardized data exploration, model building, and validation environments.

Bias and Fairness Testing Toolkits → Facilitate assessing and mitigating discriminatory outcomes.

Explainability Tools → Enable interpretation of AI models, especially for high-stakes decisions.

Model Monitoring and Performance Management → Implement systems to track model behavior and drift in production.

Risk Assessment and Mitigation

Develop a Risk Taxonomy → Categorize AI risks by domain (data, model, security, operational, etc.) and severity.

Risk Assessment Methodology → Establish processes to proactively identify, assess, and prioritize AI risks across the development lifecycle.

Mitigation Strategies → Provide templates and guidelines for addressing different risk categories.

Compliance and Regulatory Monitoring

Track Relevant Regulations → Stay current on existing and emerging AI regulations across jurisdictions.

Audit and Reporting → Establish mechanisms to ensure adherence to regulations and internal policies, providing actionable insights.

Education and Awareness

Training Programs → We offer tailored training sessions on AI principles, ethics, risk management, and regulations for developers, project managers, business stakeholders, and executives.

Communication Strategy → Communicate best practices, updates, and successes regularly to foster a culture of responsible AI throughout the organization.

Continuous Improvement

Metrics and KPIs → Define KPIs to track the CoE's performance (e.g., model fairness metrics, risk management effectiveness, policy adoption, training participation).

Feedback and Adaptation → Encourage stakeholder input and adapt policies and processes based on lessons learned.

Key Considerations

Organizational Culture → Success depends on fostering a collaborative culture where responsible AI is everyone's concern.

Adaptability → This is a rapidly evolving field. The strategy must be flexible to accommodate new regulations and technological advancements.

Communication and Transparency → Build trust through clear communication of AI initiatives, policies, and decision-making processes.

To tailor an AI Governance CoE strategy for specific industries, focus on the unique regulatory requirements, risks, and ethical considerations involved. In financial services, we prioritize model validation, explainability, and strict security to meet regulatory compliance. In technology, address data privacy, bias mitigation in customer-facing AI, and potential AI-specific regulations. Healthcare demands rigorous model testing, explainability for clinical use, and patient data protection. Across all industries, consider dedicated working groups within the CoE, external collaboration with regulatory bodies, and industry-specific risk assessments to ensure your strategy aligns with the sector's particular needs. Here are some specific details on each industry mentioned above and how to tailor your CoE for increased efficacy:

CHAPTER 7   AI GOVERNANCE AND OVERSIGHT MODEL

Financial Services

Focus Areas → Fraud detection, credit risk assessment, algorithmic trading, regulatory compliance (e.g., anti-money laundering, fair lending regulations).

Regulatory Emphasis → Adherence to strict regulations like the Fair Credit Reporting Act (FCRA), Sarbanes-Oxley Act (SOX), and the Bank Secrecy Act (BSA).

Tailoring the Strategy

Prioritize robust model validation and explainability, especially for credit-related decisions. Ensure comprehensive documentation and audit trails for compliance purposes. Emphasize all security measures to protect sensitive financial data.

Technology

Focus Areas → AI-driven software development, customer service, cybersecurity, data protection, and compliance with emerging AI-specific regulations.

Regulatory Emphasis → GDPR, CCPA, and other data privacy laws are paramount. Also, anticipate upcoming AI-specific regulations.

Tailoring the Strategy

Prioritize fairness and bias mitigation in AI-enabled products and services. Emphasize explainability and transparency, as models may have wide-reaching customer impacts. Implement robust cybersecurity measures focused on AI system vulnerabilities.

Healthcare

Focus Areas – Diagnosis and treatment recommendations, medical imaging analysis, drug discovery, precision medicine, and adherence to strict patient privacy regulations.

Regulatory Emphasis → FDA guidelines for AI-powered medical devices, HIPAA compliance, and extremely sensitive patient data protection.

CHAPTER 7   AI GOVERNANCE AND OVERSIGHT MODEL

Tailoring the Strategy

Emphasize rigorous testing and validation before deploying AI models in clinical settings. Prioritize explainability to ensure clinicians understand and trust AI-derived recommendations. Design AI systems with patient safety and well-being as the top priority.

General Considerations Across Industries

Industry-Specific Working Groups → Within the CoE, form sub-groups focusing on each industry's unique regulations and ethical concerns.

External Collaboration → Engage with industry associations and regulatory bodies to stay ahead of evolving standards.

Proactive Risk Assessment → Tailor the risk taxonomy to each industry's specific AI application scenarios.

Important note   *These are starting points. A successfully tailored CoE strategy requires deep consultation with experts in each industry to address sector-specific regulations, risks, and ethical considerations fully. Service providers like **CyberOne Security** offer consulting services to assist in setting up your CoE strategy.*

More details and information on developing your CoE can be found online and in other research documentation.

CoE Management Structure

Oversight of an AI Governance CoE often involves a combination of individuals and committees to ensure a balanced approach:

Steering Committee → A high-level group of executives from relevant business units, IT, legal, and compliance. Key responsibilities include:

- Setting strategic direction and priorities for the CoE
- Resolving cross-departmental conflicts

- Allocating resources and budgets
- Ensuring alignment between the CoE and the organization's broader goals

CoE Lead → A dedicated leader who oversees the day-to-day operations of the CoE. They typically have expertise in data science, risk management, and AI ethics. Responsibilities include

- Developing and implementing AI policies and standards
- Establishing processes for model development, validation, and deployment
- Providing training and support to teams across the organization
- Monitoring AI-related risks and reporting to the steering committee

Cross-Functional Team → The CoE draws on representatives with different specializations:

- Data scientists and AI engineers
- Legal and compliance experts
- Risk management specialists
- Business domain experts
- Ethically minded AI researchers or ethicists

External Oversight (In Highly Regulated Industries):

- Independent auditors: They may be engaged in periodic assessments of AI governance practices, especially in sectors like finance or healthcare.
- Regulatory bodies: In some cases, regulatory agencies may have oversight or the ability to audit AI systems, ensuring compliance with relevant regulations.

CHAPTER 7   AI GOVERNANCE AND OVERSIGHT MODEL

Important Considerations

Reporting Structure → The ideal reporting line for the CoE depends on the organization. It might report to the CIO, CISO, Chief Risk Officer, or the CEO.

Board-Level Involvement → Increasingly, the company's board of directors is taking an active role in overseeing AI governance to mitigate risk and ensure ethical practices.

Key AI Governance Roles and Responsibilities

Your AI governance structure should evolve with your organization. Don't hesitate to adapt the committee's composition, responsibilities, and processes as your AI initiatives become more sophisticated. Here are a few recommendations on team structure:

- **Chair** → A senior executive with authority to drive decisions and ensure accountability.

- **Data Science/AI Expertise** → Members with a deep understanding of AI model development, potential biases, and technical risks.

- **Business Representatives** → Individuals from key business areas utilizing AI to ensure alignment with business goals and identify use-case-specific risks.

- **Risk Management** → Experts in risk assessment, mitigation strategies, and compliance frameworks.

- **IT and Cybersecurity Expertise** → Individuals from the cybersecurity executive leadership team who understand data governance, AI lifecycle processes, and algorithms.

CHAPTER 7   AI GOVERNANCE AND OVERSIGHT MODEL

- **Legal & Compliance** → Members who can interpret regulations (GDPR, etc.) and ensure adherence to legal obligations related to data and AI.

- **Ethics Expertise** → Professionals who can guide ethical considerations in AI development, such as fairness, transparency, and societal impact assessment.

- **External/Independent Advisor (Optional)** → An external expert can provide an objective perspective and knowledge of industry best practices.

Chairperson. Leads and facilitates committee meetings, sets the agenda, and ensures productive discussions – acts as a liaison between the committee and senior leadership, communicating key decisions and risks. Champions AI governance initiatives throughout the organization, securing support and resources. Ensures committee recommendations are implemented and oversees adherence to AI governance policies.

Data Science/AI Expert(s). Provides technical expertise on AI model development, deployment, and monitoring. Assesses technical risks associated with AI models, such as the potential for bias, errors, or security vulnerabilities. Advises on strategies for data quality control, model validation, and model explainability. Stay up-to-date on emerging AI technologies and industry best practices.

Business Representative(s). Represents the interests of business units using AI systems. Identifies potential use cases for AI and assesses business value and alignment with strategic objectives. Communicates use-case-specific ethical and risk considerations to the committee. Ensures adherence to AI governance policies within their respective business units.

Risk Management Expert(s). Establishes a comprehensive risk assessment framework for AI initiatives. Identifies potential risks associated with AI models (reputational, operational, financial, etc.). Develops risk mitigation strategies and oversees their implementation. Monitors key risk indicators related to AI systems.

IT and Cybersecurity Experts. *The CIO* focuses on integrating AI into the organization's technological infrastructure, ensuring scalable, reliable, and efficient systems for handling AI workloads. They also drive AI adoption and innovation, aligning AI initiatives with overall business objectives. *The CISO* is responsible for securing AI systems and data, protecting against vulnerabilities, and ensuring compliance with security and privacy regulations. They work to mitigate risks associated with AI models, such as data breaches or adversarial attacks.

Legal and Compliance Expert(s). Interprets relevant regulations (GDPR, CCPA, potential future AI laws) and their implications for the organization's AI use. Ensures compliance with legal requirements regarding data privacy, intellectual property, and liability. Reviews and advises on AI-related contracts and agreements with vendors and works with the committee to develop policies addressing legal and compliance risks.

Ethics Expert (s). Guides the committee on ethical considerations related to AI development and use. Assesses AI projects for potential biases, fairness implications, and unintended societal consequences. Develops ethical guidelines and principles for the organizations' AI initiatives and promotes a culture of ethical AI development and use throughout the organization.

External/Independent Advisor (Optional). This advisor provides an objective, outside perspective on AI governance practices. They offer insights into industry trends, emerging risks, and evolving regulatory landscapes, bringing credibility and additional knowledge to the committee.

Additional Considerations

- Rotating Members → Consider rotating some members to bring fresh perspectives and expertise in different areas of AI application.

- Documentation Specialist → A dedicated person may be valuable in ensuring the well-maintained meeting minutes, policies, and decision records.

CHAPTER 7   AI GOVERNANCE AND OVERSIGHT MODEL

Committee's Responsibilities

- **Policy Development and Review** → Creating and regularly updating AI governance policies covering data, modeling, ethics, risk assessment, etc.

- **Risk Assessment and Monitoring** → Establishing a framework for identifying and quantifying AI-related risks and setting up monitoring systems

- **Project Review and Approval** → Evaluating high-risk AI projects before deployment, ensuring they adhere to policies and address ethical concerns

- **Incident Response** → Developing procedures for handling AI-related incidents, such as data breaches, model failures, or discriminatory outputs

- **Education and Awareness** → Providing training and resources for the wider organization on AI governance and responsible AI use

Committee Structure Considerations

- **Size** → Balance comprehensiveness with efficiency. Start with a core team of 6–8 members and scale as needed.

- **Reporting Lines** → Consider whether the committee reports to the board, CEO, or a designated risk/compliance officer.

- **Frequency of Meetings** → Meetings should be frequent enough to address risks (quarterly is a good start) but not overly burdensome.

- **Subcommittees** → For larger organizations, consider subcommittees focused on specific areas (ethics, cyber, data, IT, technical risk, etc).

Additional Tips

- **Balance of Authority** → Empower the committee to make decisions and enforce policies while providing appropriate escalation paths.
- **Diversity** → Ensure diverse perspectives and backgrounds are represented to mitigate bias and blind spots.
- **Documentation** → Keep thorough records of meetings, decisions, and risk assessments for accountability and auditing.

Structuring a Subcommittee

- **Membership** → Each subcommittee should consist of core governance committee members and additional subject matter experts as needed.
- **Reporting** → Subcommittees report back to the main governance committee with recommendations, raising critical issues for decision-making.
- **Oversight** → The governance committee chair should maintain oversight, ensuring subcommittee work aligns with the overall governance strategy.

Additional Tips

- **Start Small** → Begin with 1-2 subcommittees based on your organization's most pressing needs and evolve as your AI initiatives become more complex.

CHAPTER 7   AI GOVERNANCE AND OVERSIGHT MODEL

- **Cross-Subcommittee Collaboration** → Encourage communication between subcommittees to address issues from multiple perspectives.

- **Review Regularly** → Reassess the need for subcommittees periodically to ensure they remain effective.

Should I have a RACI? Yes, and here are the reasons for it! What is a RACI Model? RACI stands for:

- **R**esponsible → The person(s) actively doing the work to complete a task or decision.

- **A**ccountable → The individual is answerable for the correct and thorough completion of the task or decision, typically with the authority to make it.

- **C**onsulted → Individuals whose expertise or input is sought before a decision or task is completed.

- **I**nformed → Those who need to be kept updated about progress or the outcome of a task or decision.

The Value of a RACI Model

Clarity and Accountability → A RACI matrix clearly defines who is responsible for each aspect of AI governance (policy development, risk assessment, project approval, etc.), reducing confusion and promoting ownership.

Streamlined Decision Making → The RACI model outlines who needs to be involved and at what stage, preventing bottlenecks and delays in the governance process.

Improved Collaboration → By identifying those to be consulted, the RACI model ensures valuable input from various stakeholders is considered, fostering better decisions.

Risk Mitigation → Proactively assigning accountability reduces the risk of tasks or critical decisions falling through the cracks, mitigating potential compliance or ethical issues.

Transparency → A well-defined RACI matrix makes it clear to the wider organization, which is responsible for various aspects of AI governance, increasing visibility and trust.

RACI Management and Maintenance

Identify Key Tasks and Decisions

Start with your governance policies. Break down each policy area (like data management, model development, ethics, etc.) into specific tasks or decisions involved in their implementation. Consider project lifecycle: Include typical tasks in the AI project lifecycle, like model development, validation, deployment, and monitoring.

Define Roles

List relevant roles from your governance structure: Include members of the governance committee, subcommittees, and other key stakeholders who might be involved in AI projects.

Create the RACI Matrix

Use a table or spreadsheet to list tasks/decisions on one axis and roles on the other. For each intersection of a task and role, design a RACI designation and determine whether the role should be Responsible **(R)**, Accountable **(A)**, Consulted **(C)**, or Informed **(I)**. There should be only one person accountable (A) per task.

Review and Refine

Get feedback from stakeholders. Share the draft RACI matrix with the identified role holders to ensure accuracy and obtain buy-in. Based on the feedback from other stakeholders or participants, adjust and update the matrix.

Implement and Communicate

Formalize the RACI and incorporate it into your AI governance policies and procedures. Educate and train relevant personnel on the RACI model and their designated roles. Reference it regularly and use the RACI matrix during project planning, meetings, and decision-making processes.

Maintain and Update

Schedule regular reviews of the RACI matrix to ensure it aligns with changes in your AI landscape, governance structure, or organizational processes. Using technology, specifically project management tools and online platforms, can help manage and update your RACI matrix. Please keep it simple, start with a basic RACI structure, and expand it as your AI governance matures.

Who Owns the RACI?

There is not a single person who universally "owns" a RACI matrix. Here is how to think about ownership in the context of AI governance:

Development and Maintenance → Typically, the AI Governance Committee (or an associated subcommittee) takes primary ownership of developing and updating the RACI matrix as a key part of establishing the governance framework.

Specific Tasks/Projects → Within individual AI projects, the project manager or lead may adapt or refine a portion of the RACI matrix relevant to their tasks and team members. This ensures the RACI aligns with the project's needs.

Oversight → Ultimately, the AI governance committee chair or a designated senior leader holds overall accountability for ensuring the RACI matrix is effective, used appropriately, and updated as needed.

Collaboration → Having input from various stakeholders (governance committee, project leads) means the RACI reflects the reality of how work gets done and who has the necessary authority and expertise.

Adaptability → Distributed ownership allows the RACI to evolve with the organization and its AI initiatives instead of being a rigid, top-down document.

CHAPTER 7   AI GOVERNANCE AND OVERSIGHT MODEL

Example: Applying RACI in AI Governance

Task/Decision	Responsible	Accountable	Consulted	Informed
Develop AI ethics policy	Ethics Expert	Chair	Legal, Business Rep, Data Scientist	Senior leadership, Employees
AI project risk review	Risk Expert	Chair	Data Scientist, Business Rep, Ethics Expert	AI governance committee members
Incident response	Data Scientist	Risk Expert	IT Security, Legal, PR	Senior leadership, AI team

Figure 7-2. Sample High-Level RACI Chart

- **Align to the broader data governance policies and practices** with Canadian Tire's data privacy and protection controls, particularly the use of sensitive or otherwise risky data, *i.e., map AI model governance data governance and management procedures with Canadian Tire's enterprise data governance and privacy management processes and ensure alignment with legal's and cybersecurity's plan to protect sensitive information. In addition, ensure the AI MRG controls and risks are clearly documented and understood for proper AI controls validation and risk management.*

CHAPTER 7   AI GOVERNANCE AND OVERSIGHT MODEL

- **Outline and document risk mapping and measurement processes and standards**, i.e., *document AI modeling control procedures that map to identified risks, business processes, and regulatory requirements, if applicable. Risks should be measured using the organization's risk taxonomy, risk tolerance levels, and internal cybersecurity and data protection standards.*

- **Outline change management requirements**, *i.e., ensure that model development lifecycle processes are documented and used like your SDLC processes. Before promoting models into production, all model development and changes should undergo formal change management procedures and controls.*

- **Implement a review process for legal and risk functions**, *i.e., ensure a process exists where legal and risk teams are aware of new models being implemented, especially high-risk, high-tiered models. Some organizations have a separate review process in this instance; however, some organizations add legal and risk to special CAB sessions when the models meet certain risk and tier levels.*

- **Establish the frequency and detail for monitoring, auditing, and review processes. For example, ensure that the internal audit incorporates your model development lifecycle across the business units into their audit plans and programs and that security procedures are** *in place to monitor the activities of your models from a cyber risk and security perspective. An additional recommendation is to have a peer review of model development across the isolated business units before promoting models into production.*

- **Implement a process control to detail and test incident response plans,** *i.e., ensure your model development processes are incorporated and aligned to your security incident response plan in case of an incident or data breach due to model deficiencies or policy infractions.*

- **Establish policies for a model documentation inventory system and regularly review its completeness, usability, and efficacy,** i.e., establish and verify documentation policies for AI models are standardized across the various business units and remain current.

- **Develop and implement processes for internal and external stakeholder engagement,** *i.e., ensure a documented strategy exists for the various business units to have a VP-level signoff on all high-risk, high-tiered models being promoted to production within their business area.*

Structuring My Board of Directors

In my humble but professional experience in advising companies on what their board composition strategies should encompass, I have provided the following guidelines and approaches:

New Blood → Identify at least one, ideally two, potential board members who deeply understand AI technologies, their applications, potential biases, and the emerging regulatory landscape.

Complementary Skills → Focus on balance. Include members with risk management expertise, legal acumen (especially relating to data laws), and potentially an ethicist or social scientist focused on the wider impacts of technology.

Balance with Existing Expertise → New AI-focused members should integrate well with the board's existing business and strategic expertise from a holistic perspective.

Roles and Responsibilities

Dedicated AI Committee (Optional) → For large, AI-driven organizations, a dedicated AI oversight committee reports to the full board, ensuring focused attention.

Clear Mandate → Define the board or committee's authority. They should have insight and may approve high-risk AI projects. They should set ethical guidelines and monitor ongoing compliance requirements.

Education Is a MUST → Provide ALL board members with AI governance training tailored to their role. This is not a one-time thing but ongoing as the field evolves.

Oversight Best Practices

Information Flow → Establish how the board receives updates on AI projects, key metrics, risk assessments, and compliance audits.

Regular Risk Reviews → Mandate quarterly (or more frequent) AI-specific risk reviews covering legal, reputational, ethical, and operational risks.

External Input → Encourage the board/committee to engage with independent experts for periodic assessments and fresh perspectives on emerging risks.

Proactive Approach → The board's role is to champion responsible AI, not just police it. They should actively shape AI strategy in line with the company's values.

Additional Considerations

Board Dynamics → Fostering open communication is vital. Tech-savvy members must educate without intimidating others, ensuring the board is a cohesive team.

CHAPTER 7   AI GOVERNANCE AND OVERSIGHT MODEL

D&O Insurance → Review your Directors and Officers liability insurance, ensuring it adequately covers AI-related risks.

Start Where You Are → If a full board revamp isn't feasible, start by upskilling existing members and potentially adding an AI advisor in a non-voting capacity.

AI Governance Oversight Committee Recommendations

I will start right off by providing a few high-level recommendations based on existing client use cases.

- **Establish an AI Model Governance Oversight Committee or Council.** Implement people, processes, and controls to govern AI Models to ensure trustworthiness and compliance with applicable internal and external requirements. For example, develop an oversight council or team that understands the risks associated with your models and ensures *that Canadian Tire is on track to reap the benefits of AI while avoiding its harms and litigation risks.*

- **Define charters, playbooks, key terms, and concepts related to AI systems and the scope of their purposes and intended uses.** Document *key AI terms in your AI playbook/guides, their business goal and scope, and a support model aligning with the AI model lifecycle.*

- **Integrate AI model governance into existing organizational governance and risk committees.** AI governance and model risk management controls, processes, risks, and other attributes should be integrated with other organizational governance *and risk management committees, processes, and controls.*

CHAPTER 7 AI GOVERNANCE AND OVERSIGHT MODEL

Summary and Thoughts

Effective AI governance and oversight modeling demand a top-down approach. Securing executive and board-level buy-in is essential and paramount. AI governance is not just risk mitigation but also a strategic enabler of responsible innovation and a competitive advantage in building public trust. Educate senior leadership on the potential legal, financial, and reputational consequences of poorly governed AI to emphasize the importance of their active support.

Once leadership is on board, obtaining adequate funding is crucial. AI governance requires investment in talent, potentially including specialists in AI ethics or risk assessment. Dedicated tools for data lineage tracking, model bias testing, and continuous monitoring can improve efficiency and accuracy. Budgeting for external consultations with AI governance experts can be valuable, especially if building a comprehensive framework from scratch. Clearly articulate the potential return on this investment by highlighting reduced risks, greater agility in responding to regulations, and a stronger ethical reputation for the organization.

The board and senior management are pivotal in ensuring AI governance and risk management. They are responsible for setting the organization's ethical AI principles, allocating adequate resources for robust governance frameworks, and establishing a culture of accountability throughout the AI lifecycle. This includes mandating regular risk assessments, overseeing the implementation of policies, monitoring key metrics, and fostering open communication channels between the governance committee, technical teams, and business units. Ultimately, the board and senior management promote organizational innovation and foster an environment where responsible and effective AI governance can thrive.

Boards of directors and their structure will evolve to accommodate the complexities of AI, requiring tech-savvy members with expertise in AI ethics, risk management, and relevant regulations. Existing directors will need to upskill to avoid becoming obsolete or futile proactively. Boards should include specialized AI committees and face increased scrutiny and potential liability for AI governance. AI expertise on the board will turn governance from a risk mitigation exercise into a strategic driver of responsible innovation. Still, this evolution demands a culture of continuous learning and adaptation.

Quiz

1. What is a recommended structure for AI governance in large AI-driven organizations?

    A) A single data scientist handling all AI governance tasks

    B) A dedicated AI oversight committee reporting to the full board

    C) Outsourcing AI governance to an external agency

    D) Ignoring AI governance and focusing on development

2. Which role in the AI governance structure is responsible for ethical considerations in AI development?

    A) Data Scientist

    B) Risk and Compliance Officer

    C) AI Ethicist

    D) Business Stakeholder

CHAPTER 7   AI GOVERNANCE AND OVERSIGHT MODEL

3. What is essential for AI systems to ensure traceability and accountability throughout their lifecycle?

   A) Rapid deployment

   B) Continuous monitoring and feedback loops

   C) Minimal documentation

   D) Limited stakeholder engagement

4. Which principle is NOT typically a part of AI ethics?

   A) Fairness

   B) Transparency

   C) Profit maximization

   D) Accountability

5. True or False: AI governance and oversight are only about compliance and not about building a sustainable AI-driven future.

6. True or False: Human oversight is an essential component of AI governance, especially for high-risk models.

7. True or False: A centralized AI governance model is always more effective than a decentralized model.

8. A well-structured AI governance model should define roles and responsibilities using a _____ matrix.

CHAPTER 7   AI GOVERNANCE AND OVERSIGHT MODEL

9. Embedding robust governance practices throughout the AI lifecycle includes mandatory rigorous _____ governance.

10. Establishing an incident response plan for AI-related issues and conducting regular _____ are essential for effective AI governance.

CHAPTER 8

Managing and Addressing AI Compliance

The AI Compliance Hype and History

There was not a single defining moment that ignited the AI compliance "hype," but rather a gradual increase of attention stemming from several factors converging:

High-Profile AI Failures → As AI systems became more widely adopted, cases of bias, discrimination, and privacy violations started grabbing headlines. This fueled public concern and scrutiny of how AI is used.

Growing Regulatory Focus → The EU's landmark GDPR (effective in 2018) set a high bar for protecting individuals regarding automated decision-making. This signaled regulators' move toward more explicit AI-specific regulations.

Calls for Ethical AI → Academics, researchers, and advocacy groups amplified the discussion around ethical concerns in AI, pushing for accountability and responsible practices in this field.

Business Risk Awareness → Organizations began recognizing that AI failures could translate to reputational damage, legal liability, and financial loss, prompting them to take compliance more seriously.

Timeline of Increased Focus:

2016-2018 – Early frameworks and principles on ethical AI emerge from organizations like the Partnership on AI and the Future of Life Institute.

2019-2020 – GDPR enforcement highlights the implications for AI systems, and proposals like the EU's AI Act surface, indicating a shift toward stricter regulation.

2021-Present – AI compliance becomes a significant discussion point in businesses. Vendor solutions, consultancy firms, and specific AI actors in AI governance have begun to proliferate.

Important note *The focus on AI compliance is justified – there are very real risks associated with poorly governed AI systems. However, "hype" can sometimes lead to oversimplified solutions or a rush to demonstrate compliance rather than focusing on fundamentally addressing ethical and responsible AI development practices.*

Proposed, Enacted, and Emerging AI Regulations

Unfortunately, there is no single, comprehensive global AI regulation covering all countries and industries. AI governance is rapidly evolving, with regulations emerging or being proposed in numerous jurisdictions. Here is a breakdown of the current landscape:

EU Artificial Intelligence Act → A comprehensive risk-based approach is proposed. It classifies AI systems into "high-risk" categories with stricter requirements around transparency, bias mitigation, and human oversight. It is still under development but could set a global standard.

UK AI Regulation → The UK's proposed white paper outlines a more principles-based, less prescriptive approach to ensuring trustworthy AI. Details on specific measures are still being worked out.

US Algorithmic Accountability Act → This proposed bill (but not law) focuses on auditing AI systems for bias and impact. The United States sees more activity at the state and local levels around AI-related regulations.

Other countries, such as Canada, China, India, and many others, are developing their national AI strategies with varying degrees of focus on ethics and regulation.

Key Trends to Watch

Risk-Based Approaches → Most regulations will categorize AI systems by risk, with stricter rules for those in areas like healthcare, finance, and criminal justice.

Emphasis on Explainability and Transparency → Regulations may require the ability to explain how certain AI decisions were reached, especially for high-risk systems.

Focus on Bias and Fairness → Mitigating discrimination caused by AI is a key area of regulatory focus – not only in protected classes but also in broader societal fairness.

Tips for Navigating Compliance

Do Not Wait for Regulation to Act → Proactively adopt ethical AI principles, even without strict laws yet in force. This fosters better practices and prepares you for upcoming rules.

Monitor the Landscape Continuously → AI regulation is a rapidly evolving area. Dedicate resources to staying informed of global changes.

Do Not Focus on Just One Jurisdiction → If you operate internationally, you must comply with several regulations with varying jurisdictional or country-specific requirements for AI systems.

Major Existing Regulations

General Data Protection Regulation (GDPR – EU) → While not AI-specific, the GDPR has significant implications for AI. It mandates fairness, transparency, purpose limitation, restrictions on automated decisions, and a potential "right to explanation" for AI systems using personal data.

California Consumer Privacy Act (CCPA) and Similar State Laws → The CCPA (and similar laws in Colorado and Virginia) provides consumers with rights regarding their personal data used in AI systems, including the right to opt out of the sale of their data for targeted advertising.

Sector-Specific Regulations → In various industries, such as healthcare and finance, existing regulations (HIPAA for healthcare data, for example) become relevant when AI is used. These often affect data privacy, model accuracy, and bias prevention.

Important note *It's essential to consult legal counsel for specific guidance about regulations impacting your organization's use of AI, given the complexity of this landscape!*

Individual States Begin to Regulate AI in Absence of Federal Legislation

At the time of this writing, we see individual States begin to regulate AI without Federal Legislation. While US Lawmakers and Congress have acknowledged the necessity for federal AI legislation, progress has been sluggish. This lack of federal regulation has left a void that individual states are now stepping in to fill. This sequence of events mirrors what transpired after the European Union implemented the General Data Protection Regulation (GDPR). The GDPR established itself as a global benchmark, yet the US federal government failed to enact a comparable nationwide

CHAPTER 8   MANAGING AND ADDRESSING AI COMPLIANCE

privacy law. Consequently, states like California have taken the lead by enacting their own privacy laws, often modeled after the GDPR. This book outlines the recent developments in US state legislation regarding artificial intelligence (AI):

- **Connecticut** → A similar AI law failed to pass in Connecticut, with Governor Ned Lamont expressing his intent to veto it if passed. He argued that collaboration with other states on AI development and deployment in the private sector would be more beneficial for Connecticut.

- **Utah** → In April 2024, Utah enacted the "Artificial Intelligence Amendments," a consumer protection bill aimed at AI. This legislation includes disclosure requirements when individuals use or interact with generative AI.

- **Tennessee** → March 2024 saw the enactment of the "Ensuring Likeness, Voice, and Image Security (ELVIS) Act of 2024" in Tennessee. This act aims to combat the increasing use of AI-generated voices and deepfakes.

(Perkins Coie LLP, 2024)

Colorado: Consumer Protections in Interactions with Artificial Intelligence Systems

On May 17, 2024, Colorado enacted a new law concerning "Consumer Protections in Interactions with Artificial Intelligence Systems" (the Colorado AI Act). The law takes effect on February 1, 2026.

Similar to the EU AI Act, the Colorado AI Act adopts a risk-based approach to AI regulation by imposing certain notice, documentation, disclosure, and impact assessment requirements on developers and deployers of "high-risk" AI systems, which are defined to mean any AI system that "makes, or is a substantial factor in making, a consequential

213

decision." The Colorado AI Act defines a "consequential decision" to mean any decision that "has a material legal or similarly significant effect on the provision or denial to any consumer, or the cost or terms of" education, employment, financial, or lending services essential government services, healthcare services, housing, insurance, or legal services. As a result, the implications of the new law should be considered by a significant number of industries across the state.[2]

The Colorado AI Act focuses primarily on consumer protection, requiring developers and deployers of high-risk AI systems to use reasonable care to protect consumers from any known or reasonably foreseeable risks of algorithmic discrimination arising from the use of high-risk AI systems. The new law defines "algorithmic discrimination" to mean any condition in which the use of an AI system results in "an unlawful differential treatment or impact that disfavors an individual or group of individuals" based on their age, color, disability, ethnicity, genetic information, limited proficiency in the English language, national origin, race, religion, reproductive health, sex, veteran status, or other classification protected under Colorado or federal law.

Colorado's attorney general has exclusive authority to enforce the Colorado AI Act, which expressly excludes any private right of action. Any violation of the Colorado AI Act would constitute a deceptive trade practice. However, developers and deployers would have a rebuttable presumption if they establish compliance with the law's requirements and an affirmative defense if they take certain steps to mitigate a violation of the law.

Colorado's new law comes on the heels of an AI regulatory bill that failed to pass in Connecticut. There, the state senate passed a bill substantially similar to the Colorado AI Act, but it died in the state's house of representatives following public statements from Connecticut's governor that he would veto the bill. In Colorado, the new law faced significant opposition from industry. Although Colorado's governor signed the bill into law, he took the extra step of writing a letter to lawmakers about his reservations. In his letter, Governor Polis expressed appreciation for the bill's "interest in

CHAPTER 8   MANAGING AND ADDRESSING AI COMPLIANCE

preventing discrimination and prioritizing consumer protection," though he also emphasized that the new law should be improved upon before taking effect. For instance, while laws that seek to prevent discrimination generally focus on prohibiting intentional discriminatory conduct, this bill deviates from that practice by regulating the results of AI system use, regardless of intent; Governor Polis encouraged the Colorado legislature to reexamine this concept as the law is finalized before it takes effect in 2026.

Utah: Artificial Intelligence Policy Act

Utah's AI Policy Act went into effect on May 1, 2024. The AI Policy Act includes disclosure obligations whenever someone causes generative AI "to interact with a person" in connection with laws administered and enforced by Utah's Division of Consumer Protection. Specifically, those providing generative AI must "clearly and conspicuously" disclose that the other person is interacting with generative AI, not a human if that person asks. Further, anyone who provides services of an occupation regulated by the US Department of Commerce that requires a license or state certification must disclose that the other person is interacting with generative AI at the beginning of a verbal or written exchange.

Tennessee: The Ensuring Likeness, Voice, and Image Security (ELVIS) Act

The ELVIS Act takes effect on July 1, 2024. It expands Tennessee's right of publicity law, providing that every individual has a property right in the use of their own name, photograph, likeness, or voice. These rights are exclusive to the individual during their lifetime and for ten years after their death, with the rights then transferring to their executors, heirs, assigns, or devisees.

The ELVIS Act also creates private rights of action for unauthorized commercial use of an individual's name, photograph, likeness, or voice. Notably, it prohibits both knowingly making available a person's voice or likeness to the public without authorization and making available a technology, service, or device "the primary purpose or function of which is the production of an individual's photograph, voice, or likeness without authorization." Chancery and circuit courts may grant injunctions on terms they deem reasonable to prevent or restrain such unauthorized uses.

215

CHAPTER 8   MANAGING AND ADDRESSING AI COMPLIANCE

There are some "fair use" exceptions under the ELVIS Act for the unauthorized use of an individual's name, photograph, likeness, or voice. This includes use of those qualities in connection with a news, public affairs, or sports broadcast or account to the extent that use is protected under the First Amendment of the US Constitution.

California: Regulatory Enforcement

The California Civil Rights Council recently proposed regulations to protect against potential discrimination in hiring decisions resulting from use of AI and other automated systems.

In addition, California's Consumer Privacy Protection Agency continues to develop regulations related to automated decision-making technology, having last released a draft version at its board meeting in early March. The agency is now in the process of hosting pre-rulemaking stakeholder sessions (scheduled through May) to receive feedback on its proposals. At a high level, these draft regulations would implement a consumer's right to opt out of, and access information about, a business's use of automated decision-making technology. The draft regulations also would require a business to provide pre-use notifications to consumers, under certain circumstances, regarding how the business proposes to use automated decision-making technology.

Takeaways

Overall, AI regulations at the state level remain unsettled, with some states pushing forward with regulatory regimes tailored toward their specific concerns. In this context, Colorado's approach to AI regulation stands out as the most comprehensive and risk-based effort in the United States, resembling the EU AI Act.

Colorado's efforts initially suggest AI regulation may follow a similar path as consumer privacy laws, with proactive states taking inspiration from a comprehensive EU regulation (i.e., the GDPR for privacy protection and the EU AI Act for AI regulation), while the US Congress fails to act. On the other hand, Connecticut's failure to pass a similar bill inspired by the EU AI Act suggests that perhaps not all states will follow Colorado's lead.

The mixed record of the state legislatures occurs amid the backdrop of little federal level activity aside from the executive branch, wherein the White House issued an expansive Executive Order last fall concerning the safe, secure, and trustworthy development and use of AI, which federal agencies have begun to implement.

Ultimately, states may face challenges passing comprehensive AI regulations due to AI developers' concerns about prematurely regulating a nascent industry and potentially stifling innovation. Either way, the AI regulation in the United States is relatively unsettled compared to the European Union and likely will remain so in the absence of federal legislation specifically addressing AI.

(Perkins Coie LLP, 2024)

AI Compliance: The Evolving Landscape

AI compliance is an emerging field focused on ensuring that artificial intelligence systems adhere to existing laws, regulations, and ethical principles. AI's wide-reaching impact means compliance efforts need to address many areas, including

Data Privacy and Protection → AI systems often rely on vast amounts of personal data. Compliance entails following regulations like the GDPR (General Data Protection Regulation), CCPA (California Consumer Privacy Act), and industry-specific data collection, storage, usage, and consent laws.

Algorithmic Fairness and Bias Prevention → AI models inherit biases in their training data. Compliance aims to eliminate discrimination based on race, gender, age, or other protected attributes. It involves proactive bias testing and implementing mitigation strategies.

CHAPTER 8 MANAGING AND ADDRESSING AI COMPLIANCE

Transparency and Explainability → Many AI systems, especially deep learning models, operate as "black boxes." Compliance in high-stakes areas (e.g., healthcare, finance) often requires explainability – understanding how models reach decisions and providing clear justifications.

Security and Robustness → Protecting AI systems from cyberattacks and ensuring they function reliably is essential to prevent unintended consequences. Compliance involves implementing robust security measures and testing models for resilience to adversarial attacks.

Intellectual Property → AI development raises complex questions about ownership of data, software, and inventions generated by AI systems. Regulations around patents and copyrights are still evolving in this space.

Accountability and Liability → Legal frameworks are adapted to address who is liable when AI systems cause harm. Compliance involves clear documentation, audit trails, and insurance to cover potential risks.

Ensure that the documented and intended purpose, potentially beneficial uses, context-specific laws, norms and expectations, and perspective settings in which the AI models will be deployed are understood and that the model respects the privacy of its users.

The Regulatory Patchwork

Unlike established areas like data privacy, no comprehensive regulatory framework for AI exists globally. Instead, a patchwork of approaches is emerging:

Sector-Specific Regulations → Existing laws in finance, healthcare, and other sectors are being updated to address AI-specific uses. For example, the FDA provides guidelines for AI in medical devices.

Broader Non-binding Principles → Many organizations, including governments, have released high-level AI ethics principles. These help guide responsible AI development but lack legal enforceability.

AI-Specific Legislation → The EU's AI Act is the most notable example. It classifies AI systems by risk, imposing stricter requirements on high-risk applications. Other countries are exploring similar legislation.

Challenges and Best Practices for Organizations

Navigating Complexity → Staying updated on evolving regulations across different sectors and jurisdictions is a significant challenge.

Proactive Approach → Organizations can't wait for fully-fledged regulations. Best practices include establishing internal AI governance frameworks, integrating ethics into development, and ongoing risk monitoring.

Cross-Functional Collaboration → AI compliance requires input from legal, technical, risk management, and business teams to effectively address the multifaceted issues it raises.

The Future of AI Compliance

AI compliance will remain a rapidly evolving field. Organizations prioritizing ethical and responsible AI while staying agile in the face of new regulations will be best positioned to mitigate risks and leverage AI's transformational potential.

So, how do I address AI model compliance and ensure adherence?

Practical Approach to Achieve AI Compliance

In practice, I typically advise my clients to adopt a compliance framework focusing on risk-based compliance *(meaning prioritize where feasible)* and have a detailed and tailored strategy for addressing proactive AI regulatory compliance. Here is a simple four-phase approach for addressing compliance.

CHAPTER 8   MANAGING AND ADDRESSING AI COMPLIANCE

Phase 1: Knowledge and Preparation

Regulatory Landscape Mapping → Conduct a thorough analysis of relevant regulations (GDPR, CCPA, etc.) and industry-specific standards applicable to your organization and AI use cases. Create a continually updated repository of obligations.

Risk Assessment Framework → Develop a tailored risk assessment framework specific to AI systems. Identify potential areas of non-compliance, data vulnerabilities, ethical breaches, and discriminatory risks within your AI models.

Cross-Functional Team → Assemble a team beyond the AI Governance Committee. Include IT, legal, risk management, and, importantly, representatives from business units where AI is actively used.

Phase 2: Building Compliance into AI Processes

Privacy by Design → Integrate privacy principles from the design phase of AI systems. This involves data minimization, strong anonymization techniques, and user data control.

Ethical Impact Assessments → Mandate ethical impact assessments for high-risk AI projects. Employ methodologies to examine the potential for bias, societal harm, and unintended consequences and record outcomes and mitigation plans.

Explainability and Transparency → Prioritize explainability in model design depending on the use case and regulations – document data sources, modeling choices, and decision-making logic.

Vendor and Partner Management → Establish clear expectations for AI vendors regarding compliance, ethical AI practices, and data security. Incorporate these provisions into contracts.

Phase 3: Operationalizing Compliance

AI Governance Policies → Develop and regularly update comprehensive policies covering data handling, model development, risk mitigation, bias prevention, incident response, etc. Align policies with your risk assessment framework.

Auditing and Monitoring → Implement regular audits of AI models, both technical and ethical. Monitor for performance degradation, concept drift, and changing data distributions that could impact compliance and fairness.

Incident Response Plan → Create a detailed response plan for potential AI-related incidents, such as discrimination events, security breaches, or significant model failures. Include communication protocols and escalation paths.

Training and Awareness → Provide mandatory training to all relevant employees on AI compliance, responsible AI principles, and their role in identifying risks. Tailor training to specific departments.

Phase 4: Continuous Improvement and Adaptation

Feedback Loops → Establish mechanisms to gather feedback from employees, customers, and stakeholders on the effectiveness of your AI governance and the potential impact of your AI systems.

Evolving Regulations → Dedicate resources to monitoring the regulatory landscape. Proactively adapt your compliance strategy as new regulations emerge or existing ones are updated.

Collaboration and Sharing → Participate in industry forums and share best practices with peer organizations (while respecting competitive boundaries). This helps address sector-wide compliance challenges.

Additional Considerations

Documentation → For transparency and auditability, keep thorough records of compliance activities, risk assessments, model development processes, and policy updates.

Independent Oversight → Consider involving an external advisor for periodic reviews or complex ethical assessments, lending credibility to your governance process.

CHAPTER 8   MANAGING AND ADDRESSING AI COMPLIANCE

Preventing AI Compliance Surprises and "Gotchas"

Here's a breakdown of some "surprise gotchas" that can trip up even well-intentioned organizations regarding AI compliance.

Gotchas Related to Data

Data Drift → Real-world data changes over time, but models trained on historical data may become inaccurate or biased without retraining. Unexpected data drift can lead to non-compliance or unfair outcomes.

Data Provenance and Lineage → It might not be easy to track where and how your AI training data was obtained, especially in larger organizations. This risks using data obtained unethically or in violation of regulations.

"Black Box" Data Sources → Using third-party data sources or pre-trained models can introduce hidden biases or compliance risks if you don't have full visibility into how that data was collected and processed.

Gotchas Related to Models

Algorithmic Bias Amplification → Even seemingly unbiased data can contain hidden biases that models amplify. Regular fairness testing is not just about compliance; it's about preventing harm unintentionally caused by your AI systems.

Explainability vs. Accuracy Trade-offs → Extraordinarily complex models (like deep neural networks) can outperform simpler ones, but often at the cost of interpretability. This can clash with compliance and "right to explanation" requirements.

"Clever" Adversaries → Adversaries can manipulate models by feeding them specially crafted data to trigger errors or discriminatory behavior. Compliance needs to consider security risks deliberately targeting AI systems.

Gotchas Related to Deployment and Operations

Scope Creep → It is tempting to redeploy existing models for new purposes. However, a model deemed compliant in one context might have surprising ethical or legal issues when applied elsewhere.

CHAPTER 8   MANAGING AND ADDRESSING AI COMPLIANCE

Unforeseen Interactions → *AI systems integrated into complex workflows can trigger unexpected cascading effects. A compliance problem might originate from an unrelated change elsewhere in the system.*

Human-in-the-Loop Oversight → *Humans tasked with overseeing AI might over-trust automated recommendations or fail to understand the system's limitations, creating a new compliance risk.*

Mitigation Strategies

Proactive Monitoring → *Do not assess for compliance alone before launch. Systems for detecting drift, bias, and bizarre behavior are vital and must be implemented to monitor and manage mitigation continuously.*

Iterative Risk Assessments → *Include AI compliance in your regular risk assessments. As models and data evolve, so do the potential risks.*

Treat AI as its Ecosystem → *Your AI systems deserve dedicated governance processes, not simply shoehorned into existing compliance frameworks.*

Authority to Enforce Compliance → *Develop processes to assess AI actors' independence, stature, and oversight to ensure they have the autonomy and resources to perform assurance, compliance, and feedback tasks effectively. That is, define and document acceptable levels of performance differences per established organizational governance policies, business requirements, regulatory compliance, legal frameworks, and ethical standards within the context of use.*

Establish and Implement → *Establish and implement measurable baselines for sustainable AI model operation in accordance with organizational standards and policies, i.e., regulatory compliance, legal frameworks, and environmental protection and sustainability norms.*

Assess Trade-offs → *Develop and implement procedures to assess trade-offs between AI model performance and sustainable operations in accordance with organizational principles and policies, regulatory compliance, legal frameworks, and environmental protection and sustainability norms, i.e., implement risk-based compliance procedures that prioritize risks involving physical safety, legal liabilities, regulatory compliance, and negative impacts on individuals, groups, or society.*

Policies and Standards → Establish policies that incentivize AI actors to collaborate with existing legal, oversight, compliance, or enterprise risk functions in their AI model risk management activities, i.e., these policies require including oversight functions (legal, compliance, risk management) from the outset of the AI model design process.

Standards and Compliance Mapping → Develop and implement detailed standards for experimental design, data quality, and model training. Implement a process delineating experimental design from staging or development for production models. Experimental modeling should be isolated, and production data should be excluded unless data is de-identified or encrypted.

Remember, the AI regulatory landscape is still evolving. Staying on top of "gotchas" requires dedicating resources to monitoring updates and seeking expert advice when in doubt.

Balancing Innovation and AI Compliance

Striking a balance between AI innovation and compliance constantly challenges organizations. Here is a breakdown of how you can navigate this tension:

Shifting From a "Compliance Only" Mindset

Proactive Ethical Design → Go beyond mere compliance checklists. Integrate ethical considerations, privacy, and fairness into the AI design process from the very beginning. This reduces the need for costly redesigns later and fosters trust in your AI systems.

Compliance as an Innovation Driver → Reframe regulations as restrictions and prompts for creative problem-solving and the development of more robust, explainable AI systems.

Strategies for Balancing Innovation and Compliance

Federated Learning and Privacy-Enhancing Techniques → Explore methods for training models on decentralized data without compromising user privacy. This opens doors for new products while respecting regulations.

Synthetic Data → Use synthetic data for model training, especially when real-world sensitive data poses compliance hurdles. This can accelerate development and testing.

Cross-functional Collaboration → Break down silos between data scientists, legal, risk, and business teams. Early and ongoing collaboration ensures compliance is not an afterthought but baked into innovation efforts.

Explainability Tools → Invest in research and development of model explainability techniques, which will enable compliance with "right to explanation" rules without sacrificing model performance.

Testbeds and Sandboxes → Create internal test environments where new AI concepts can be evaluated for compliance and fairness without risking real-world impact.

Changing the Organizational Culture

Responsible AI Education → Train employees on AI ethics and compliance across the organization. This is for technical teams and everyone interacting with AI-powered systems.

Incentivize Ethical Innovation → Reward teams for finding creative solutions that uphold compliance and deliver business value. Celebrate successes in building trustworthy AI.

Embrace Ongoing Assessment → Do not view compliance as a one-time goal. Regularly audit models, re-assess risks as regulations evolve, and be willing to adapt your AI systems.

CHAPTER 8   MANAGING AND ADDRESSING AI COMPLIANCE

Key Points to Remember

- This is a journey, not a destination. The AI landscape and regulatory environment constantly shift, so adaptability is key.

- Communication is critical → Transparency about AI projects builds trust internally with employees and externally with customers and regulators.

- Do not let perfect be the enemy of good → Strive for continuous progress. Aim for responsible, compliant AI that delivers value, even if perfect solutions are an ongoing quest.

Summary and Key Takeaway → Effective AI compliance is not a checkbox activity, but an ongoing process interwoven with responsible AI development. It requires a culture of vigilance and adaptation across the entire organization.

Summary and Thoughts

Effectively managing AI model compliance requires a proactive, multi-faceted approach. Start by thoroughly understanding the regulatory landscape relevant to your industry and the specific use cases for your AI models. Do not exclusively focus on current regulations like GDPR; track other emerging AI-specific legislation and proactively incorporate those requirements into your AI development practices. This ensures your organization is prepared for tightening AI-based compliance requirements.

Implement a rigorous compliance framework within your AI lifecycle. This includes defining clear compliance metrics, incorporating compliance testing throughout model development, and maintaining detailed documentation of data sources, modeling choices, validation

CHAPTER 8 MANAGING AND ADDRESSING AI COMPLIANCE

results, and any compliance-related decision-making. Invest in ongoing monitoring tools to detect model drift, bias, and potential violations. Foster a culture of collaboration between technical, legal, and compliance teams to ensure alignment and address issues swiftly.

Quiz

1. What is the primary reason for the evolving landscape of AI compliance?

    A) Decreasing interest in AI technologies

    B) Increasing regulatory scrutiny and proposed AI-specific regulations

    C) Reduced need for ethical considerations in AI

    D) Lack of advancements in AI

2. Which approach is recommended for achieving AI compliance?

    A) Ignoring external regulations

    B) Integrating AI compliance with broader enterprise compliance frameworks

    C) Relying solely on internal guidelines

    D) Avoiding stakeholder engagement

CHAPTER 8   MANAGING AND ADDRESSING AI COMPLIANCE

3. What should organizations do to prevent AI compliance surprises?

    A) Conduct regular AI risk assessments and audits

    B) Avoid updating their compliance frameworks

    C) Disregard regulatory developments

    D) Minimize documentation

4. Which factor is NOT a driver of AI compliance?

    A) Regulatory requirements

    B) Organizational ethical standards

    C) Stakeholder expectations

    D) Ignoring AI risks

5. True or False: AI compliance only focuses on adhering to current laws and does not consider ethical implications.

6. True or False: Integrating AI compliance with enterprise compliance frameworks enhances overall governance effectiveness.

7. True or False: Preventing AI compliance surprises requires organizations to stay updated on regulatory changes and conduct regular risk assessments.

8. AI compliance is an evolving landscape driven by increasing regulatory scrutiny and proposed _____ regulations.

CHAPTER 8   MANAGING AND ADDRESSING AI COMPLIANCE

9. To achieve AI compliance, integrate it with broader enterprise compliance frameworks and conduct regular _____ and audits.

10. Preventing AI compliance surprises involves staying updated on regulatory changes and proactively addressing potential _____.

CHAPTER 9

Integrating AI Governance with Enterprise Governance Risk and Compliance

My extensive enterprise governance, risk, and compliance (EGRC) background inspired me to dedicate an entire chapter to this crucial topic. I have consistently championed integrated GRC approaches, helping organizations unlock the potential of shared workflows, risk assessments, controls, and governance procedures. This shift away from siloed practices has consistently yielded significant client benefits.

Integrated GRC offers significant value to organizations by breaking down silos and providing a holistic approach to managing risks, and ensuring adherence to regulations and internal policies.

Note: Because there are many publications and guides on EGRC, this book does not encompass a comprehensive review of EGRC or its Principled Performance Model. More information on EGRC can be found at www.oceg.org.

CHAPTER 9 INTEGRATING AI GOVERNANCE WITH ENTERPRISE GOVERNANCE RISK AND COMPLIANCE

What Is Enterprise Governance Risk and Compliance (EGRC)

Enterprise Governance, Risk, and Compliance Management (EGRC) is a holistic framework that guides how organizations manage a wide range of risks and ensure they operate by internal and external policies.

Components of EGRC

Governance → The overall structure of how decisions are made within an organization. This includes setting strategic direction, ethics, defining roles and responsibilities, and the board of directors' oversight role.

Risk Management → Identifying, assessing, prioritizing, and mitigating various organizational risks. These include

- Operational risks (e.g., process failures, technology breakdowns)
- Financial risks (e.g., market volatility, fraud)
- Strategic risks (e.g., competitor moves, changing consumer preferences)
- Reputational risks (e.g., scandals, ethical breaches)
- Compliance risks (e.g., violating laws or regulations specific to your industry)

Compliance → The processes and practices to ensure the organization adheres to external regulations (like industry-specific or data privacy laws like GDPR) and internal policies.

Why EGRC Matters

Integrated Approach → It breaks down silos, fostering collaboration between different departments (legal, finance, IT, etc.) and ensures a consistent approach to managing risk and compliance.

Efficiency → EGRC streamlines processes and can often leverage technological tools to reduce redundancies and improve accuracy.

Proactive vs. Reactive → Focuses on anticipating risks and compliance issues rather than just reacting to incidents, protecting the organization's bottom line and reputation.

Agility → A robust EGRC framework helps organizations quickly adapt to changing regulations and market conditions.

Stakeholder Trust → Demonstrating a commitment to EGRC reassures investors, customers, and partners, enhancing the organization's overall reputation.

AI Governance is increasingly seen as a key pillar within EGRC frameworks due to the unique risks and ethical considerations associated with AI.

Integrating AI Governance and Risk Management with the Broader Enterprise Governance, Risk, and Compliance (GRC) Framework

Integrating AI governance and risk management within your GRC framework offers numerous benefits. By aligning AI initiatives with established risk assessment protocols, compliance standards, and ethical protocols, you can effectively mitigate AI-specific risks (like bias or security vulnerabilities) while leveraging shared resources and workflows. This integrated approach fosters data consistency, reduces redundancy, and promotes a comprehensive understanding of interconnected risks across the organization. This strategy enables proactive governance, driving responsible AI innovation while protecting your organization's reputation and ensuring adherence to evolving regulations.

Successful integration is an ongoing process. Stay updated on best practices, encourage team feedback, and be prepared to adapt your approach as newer, more advanced AI technologies and regulations emerge. As documented below, you can integrate AI governance with your EGRC ecosystem.

Aligning Frameworks and Policies

Mapping → Review your existing GRC policies and identify gaps related to AI risks (e.g., bias, explainability, data protection, cybersecurity).

Updates → Update policies to include AI-specific requirements, covering the entire model lifecycle from development and testing to deployment and monitoring.

Harmonization → Ensure consistency between AI governance policies and overarching data management, risk assessment, and ethical conduct guidelines.

Adapting Risk Assessment Processes

AI-Specific Risk Taxonomy → Expand your risk taxonomy to categorize new risks inherent in AI systems, such as algorithm bias, model vulnerability to adversarial attacks, and unintended consequences in the production environment.

Assessment Methodology → Update risk assessment methodology to incorporate AI risks across various stages of the model's lifecycle. Engage AI experts and domain specialists in this process.

Leveraging GRC Tools and Platforms

Centralized Repository → Use your existing GRC platform or invest in one that can manage AI and enterprise risks. This enables a holistic view.

Workflows → Establish workflows within your GRC platform for AI project assessment, risk classification, approval processes, and ongoing monitoring.

Reporting and Dashboards → Implement dashboards that provide visibility into key AI risks, model performance metrics, and compliance status, enabling real-time tracking.

CHAPTER 9   INTEGRATING AI GOVERNANCE WITH ENTERPRISE GOVERNANCE RISK AND COMPLIANCE

Note *I typically remain agnostic about tools and solutions; however, IBM OpenPages is one of the most effective tools in the market from a GRC perspective for covering AI risk.*

Cross-Functional Collaboration

GRC and AI Teams → Establish regular communication channels between your GRC team, data scientists, and AI engineers. Foster a culture of shared responsibility.

Governance Structure → Integrate AI representatives into your GRC steering committee or create a dedicated AI working group that feeds into the main GRC body.

Joint Training → Offer cross-training for GRC personnel on AI risks and for AI teams on governance and compliance principles.

Emphasize Continuous Learning and Improvement

Feedback Loops → Establish mechanisms to learn from incidents and near misses related to AI and update policies and risk assessments accordingly.

Stay Alert for External Changes → The regulatory landscape around AI is evolving rapidly. Track developments and their implications for your enterprise GRC frameworks.

Additional Considerations

Vendor Risk Management → If you utilize third-party AI solutions, extend your due diligence and risk assessment processes to include their governance practices.

Executive Buy-in → Dedicated support from top leadership is essential for successfully integrating AI governance into your broader GRC strategy.

Let's dive into specific areas for detailed AI governance and risk management integration into your EGRC framework.

Detailed Integration Areas

Policy Alignment and Cross-Referencing

CHAPTER 9 INTEGRATING AI GOVERNANCE WITH ENTERPRISE GOVERNANCE RISK AND COMPLIANCE

Existing GRC Policies → Review core GRC policies (data privacy, cybersecurity, ethics, etc.) and explicitly incorporate AI governance principles where relevant.

AI Policy Cross-Referencing → Ensure AI policies reference relevant GRC policies, creating a cohesive framework for managing intertwined risks.

Example: Your AI Fairness policy might reference your organization's anti-discrimination policies and legal compliance objectives.

Risk Assessment and Mitigation Strategies

Tailored Risk Scenarios → Develop scenarios specific to deploying AI models, factoring in potential biases, security breaches, system failures, and unintended consequences.

Involve AI Experts → Include data scientists, AI engineers, and subject matter experts in risk assessment workshops, fostering collaboration throughout risk analysis.

Mitigation Playbooks → Create templates for mitigation strategies addressing AI-specific risks (e.g., bias testing protocols and adversarial attack response plans). Integrate these into overall incident response plans.

GRC Tooling and Workflows

Model Inventory → Maintain a centralized inventory of all AI models in use. Include details like purpose, data sources, risk classification, owners, and relevant policies.

AI Risk Scoring → Develop a risk scoring model for AI projects, considering factors like model complexity, data sensitivity, and potential impact of failure.

Pre-Deployment Checklists → Mandate the use of checklists for model development and deployment, ensuring adherence to governance policies and risk mitigation measures.

Training and Awareness

Targeted Training Modules → Design role-specific training on AI risks and governance best practices for GRC staff, IT teams, and business stakeholders involved in AI initiatives.

Collaboration Workshops → Hold joint workshops to bridge the knowledge gap between GRC and AI teams and foster a shared understanding of risks and responsibilities.

Communication Campaigns → Raise organization-wide awareness of the importance of responsible AI and the need for collaboration in managing potential risks.

Monitoring and Continuous Improvement

KPI Integration → Integrate AI-related KPIs (e.g., model fairness metrics and explainability scores) into your GRC reporting dashboards.

Periodic Audits → Establish regular audits of AI model performance, bias checks, and compliance with governance policies.

Remediation Workflows → Define clear workflows for responding to incidents arising from AI model failures, breaches, or ethical concerns.

Let's explore the integration process for the following specific risk categories in Financial Services and Healthcare:

- Data protection and governance
- Cybersecurity
- Operational risks

Data Protection and Governance

Policy Alignment → Ensure AI policies address data privacy regulations (GDPR, CCPA, etc.) and industry-specific standards for sensitive data in finance and healthcare.

Risk Assessments → Focus on potential privacy violations due to AI models (e.g., re-identification from anonymized data, biases leading to unfair treatment). Integrate data lineage tracking and privacy impact assessments into AI development.

GRC Tools → Utilize data governance features within your GRC platform. Manage AI model access controls, track consent and data usage, and integrate with data breach response plans.

Training → Emphasize privacy principles by design and responsible data handling for data scientists and developers alongside GRC personnel.

Cybersecurity

Policy Updates → Extend cybersecurity policies to address vulnerabilities specific to AI models (adversarial attacks, model poisoning).

Risk Assessments → Conduct threat modeling and risk assessments considering AI-specific attack vectors. Build robustness testing into model development and track security metrics post-deployment.

GRC Tools → Integrate AI model security monitoring into incident detection and response systems. Consider tools specialized in identifying adversarial attacks.

Training → Implement cybersecurity awareness for AI teams focused on secure coding practices and recognizing attack patterns targeting AI systems.

Operational Risks in Financial Services and Healthcare

Financial Services

Policy Considerations → Align AI governance with regulatory compliance in lending, algorithmic trading, and anti-money laundering. Emphasize model explainability and fairness throughout the model lifecycle.

Risk Assessments → Focus on scenarios involving model instability, unintended market impact, and potential compliance violations arising from AI implementations.

GRC Tools → Use your GRC platform's model tracking, reporting, and change-management features to ensure audit and regulatory requirements compliance.

Training → Educate GRC and risk management teams about financial regulations relevant to AI systems.

Healthcare

Policy Considerations → Prioritize patient safety, confidentiality (HIPAA), and explainability in AI governance, particularly for medical diagnosis and treatment applications.

Risk Assessments → Identify risks related to diagnostic errors, privacy breaches, and patient harm associated with AI systems. Closely evaluate the clinical and health impacts of AI decision-making models.

GRC Tools → To ensure patient safety, utilize features tailored to patient data sensitivity and track potential incidents related to medical AI models.

Training → Highlight the ethical use of health data and potential biases that could lead to adverse health outcomes for GRC, AI development teams, and healthcare professionals.

Additional Considerations

Third-Party Vendors → Assess the AI governance practices of vendors providing AI solutions in financial services and healthcare.

Cross-Industry Collaboration → Actively participate in industry forums to share best practices and stay ahead of emerging risks and regulations specific to AI in these sectors.

Integrating AI Risk Management with Enterprise Risk Management and eGRC

AI model risk management must become integral to your overall ERM and EGRC framework. Start by expanding your risk taxonomy to explicitly include AI-specific risks like bias, explainability issues, data quality problems, security vulnerabilities, and unforeseen operational impacts. Then, adapt enterprise risk assessment processes to proactively identify, evaluate, and prioritize the potential consequences of AI model failures on financial, reputational, or compliance levels. This necessitates collaboration between ERM, data science, and business domain experts for holistic assessments.

Aligning AI model risk management policies and controls with broader ERM principles is essential. Embed pre-deployment checklists, ongoing monitoring, and regular auditing into your governance model. Leverage GRC platforms to centralize AI model inventories, track risk classifications, automate risk reporting, and create dashboards for real-time visibility into AI-related risks across the enterprise. This integration will enable informed decision-making, effective mitigation strategies, and a robust defense against the evolving risks posed by AI adoption.

Auditing AI Controls and Processes to Validate Compliance

An AI risk management audit aims to assess the effectiveness of governance, controls, and processes in place to identify, mitigate, and monitor risks associated with AI models.

Key Areas to Focus On:

Governance and Oversight → Evaluate the presence of clear AI governance policies, defined roles and responsibilities, and effective oversight mechanisms.

Model Development and Validation → Assess the rigor of development processes, data quality checks, bias testing, validation methodologies, and documentation standards.

Deployment and Monitoring → Examine change management controls, procedures for model performance monitoring, drift detection, and incident response plans.

Security and Data Protection → Audit adherence to cybersecurity standards, data privacy regulations, and practices to protect AI models and sensitive data.

Ethical Considerations → Assess processes for identifying and mitigating potential bias and fairness issues and ensuring alignment with ethical principles.

CHAPTER 9 INTEGRATING AI GOVERNANCE WITH ENTERPRISE GOVERNANCE RISK AND COMPLIANCE

Regulatory Compliance → Verify adherence to industry-specific regulations and relevant AI-related laws.

Sample AI Risk Management Audit Program

Planning

- **Define Scope** → Determine which AI models or applications are in scope for the audit.
- **Risk Assessment** → Identify key risks associated with the selected AI models.
- **Develop an Audit Plan** → Create a plan outlining specific audit procedures, timelines, and resources.

Fieldwork

- **Interviews** → Conduct interviews with AI model developers, project managers, GRC personnel, and relevant stakeholders.
- **Review Documentation** → Examine AI governance policies, model development procedures, risk assessments, monitoring reports, and compliance documentation.
- **Technical Testing** → In collaboration with data scientists or technical experts, perform tests to assess model performance, bias, security vulnerabilities, and robustness.

Reporting

- **Document Findings** → Summarize findings on the effectiveness of AI risk management processes, identifying gaps, weaknesses, or areas for improvement.

- **Recommendations** → Provide actionable recommendations to strengthen controls, mitigate risks, and enhance AI governance.

- **Communication** → Present audit results to the AI governance steering committee, GRC team, and relevant executives.

Important Considerations

Tailor the Program → Adapt the audit scope and procedures based on the specific AI models, their use cases, and the organization's risk appetite.

Collaborative Approach → The audit should cooperate between auditors, data scientists, and the GRC team.

Regular Frequency → Schedule regular AI risk management audits to ensure ongoing governance effectiveness.

Here's a comprehensive guide on auditing AI risk and controls to ensure and validate compliance:

Planning and Preparation

- **Understand the Regulatory Landscape** → Research your industry's relevant regulations and standards and the specific AI applications you are auditing. This includes general data privacy laws (GDPR, CCPA), sector-specific regulations (healthcare, finance), and emerging AI-focused laws.

- **Align with Governance Framework** → Review your organization's AI governance policies and ensure the audit plan aligns with the defined risk management procedures.

- **Develop Audit Checklist** → Create a detailed checklist covering data governance, model bias and fairness, explainability, security, ethical considerations, and regulatory adherence specific to the AI models in scope.

CHAPTER 9 INTEGRATING AI GOVERNANCE WITH ENTERPRISE GOVERNANCE RISK AND COMPLIANCE

Audit Execution

- **Process Review and Interviews** → Examine documented processes around AI development, testing, deployment, monitoring, and incident response. Interview data scientists, risk managers, and compliance personnel to understand how these processes are implemented in practice.

- **Technical Testing** → Collaborate with data scientists and IT security teams to perform technical tests on your AI models. This might entail

- **Bias and Fairness Testing** → Use metrics and tools to assess if models produce discriminatory outputs.

- **Explainability Analysis** → Evaluate if model decisions can be explained, especially for high-stakes scenarios.

- **Security Testing** → Conduct penetration tests and vulnerability scans to identify potential attack vectors in the AI systems.

- **Documentation Review** → Scrutinize model development records, testing reports, risk assessments, change management logs, and compliance reporting for thoroughness and policy adherence.

Compliance Validation

- **Map Regulatory Requirements** → Create a detailed map between specific regulatory requirements and the controls implemented in your AI governance framework.

- **Gap Analysis** → Identify gaps where controls are insufficient or missing to address specific regulatory mandates.

- **Evidence Gathering** → Collect evidence demonstrating compliance, such as model test results, audit trails for decision-making, and security logs.

Reporting and Recommendations

- **Findings and Recommendations** → Create a clear audit report summarizing your findings, identifying strengths, weaknesses, and compliance deficiencies. Provide actionable recommendations to remediate gaps and further enhance controls.

- **Prioritization** → Help the organization prioritize risks based on the potential impact on compliance and overall business objectives.

- **Communication to Stakeholders** → Present the audit report to the AI governance steering committee, executives, and relevant stakeholders.

Additional Considerations

- **Independent Audits** → For highly regulated industries, consider engaging independent third-party auditors to increase objectivity and validate the effectiveness of your AI risk management and compliance practices.

- **Evolving Landscape** → Stay informed about changes in regulations and AI governance best practices and adapt your audit procedures accordingly.

- **Continuous Improvement** → View auditing as a tool for ongoing improvement. Use the findings to strengthen your AI risk management processes and maintain robust compliance over time.

CHAPTER 9 INTEGRATING AI GOVERNANCE WITH ENTERPRISE GOVERNANCE RISK AND COMPLIANCE

Summary and Thoughts

AI governance, encompassing ethical principles, risk management, and compliance specific to AI systems, is becoming a critical component within a comprehensive Enterprise Governance, Risk, and Compliance (EGRC) framework. Tightly integrating AI governance with existing EGRC structures ensures that organizations manage the unique risks associated with AI technology, like bias, lack of transparency, and evolving regulations. This integration enables a holistic risk identification and mitigation approach across the organization.

By incorporating AI governance within EGRC, organizations benefit from streamlined processes, reduced redundancies, and centralized oversight. This integration allows for better alignment of AI initiatives with the company's overall risk tolerance, compliance obligations, and strategic goals. Additionally, integrated EGRC platforms often offer specialized tools and analytics for tracking AI-specific metrics, improving decision-making, and demonstrating a robust governance approach to internal and external stakeholders.

Quiz

1. What is the primary benefit of integrating AI governance within the broader GRC framework?

    A) Increased operational costs

    B) Enhanced data inconsistency

    C) Proactive governance and risk mitigation

    D) Reduced need for regulatory compliance

CHAPTER 9 INTEGRATING AI GOVERNANCE WITH ENTERPRISE GOVERNANCE RISK AND COMPLIANCE

2. Which risk category does NOT typically fall under Enterprise Governance, Risk, and Compliance (eGRC)?

    A) Operational risks

    B) Financial risks

    C) Compliance risks

    D) Personal preferences risks

3. How can AI governance benefit from leveraging GRC tools and platforms?

    A) By reducing transparency

    B) By creating data silos

    C) By providing a holistic view of risks

    D) By ignoring stakeholder input

4. What is a key component of AI-specific risk assessment?

    A) Ignoring algorithm bias

    B) Model vulnerability to adversarial attacks

    C) Reducing model transparency

    D) Disregarding data protection

5. True or False: AI governance should be integrated into the broader GRC framework to ensure consistency and comprehensive risk management.

6. True or False: Integrating AI governance with enterprise risk management is a one-time process.

CHAPTER 9 INTEGRATING AI GOVERNANCE WITH ENTERPRISE GOVERNANCE RISK AND COMPLIANCE

7. True or False: Updating risk assessment methodology to incorporate AI risks helps in identifying potential issues early.

8. A robust eGRC framework helps organizations quickly adapt to changing regulations and market conditions, demonstrating a commitment to _____.

9. Enterprise Governance, Risk, and Compliance Management (eGRC) is a holistic framework guiding how organizations manage a wide range of risks and ensure they operate by _____.

10. Aligning AI initiatives with established risk assessment protocols, compliance standards, and ethical protocols effectively mitigates AI-specific risks while leveraging shared _____.

CHAPTER 10

AI Policy Management and Enforcement

AI policy management and enforcement involves creating, communicating, and actively upholding organizational rules governing AI development and use to mitigate risk, ensure compliance, and foster responsible practices.

What Is AI Policy Management

AI policy management creates, implements, monitors, and updates the rules governing AI development and use within an organization. AI policies must align with the organization's ethical principles, risk tolerance, operational goals, and relevant regulations. These policies encompass data governance, bias mitigation, explainability, model validation, and incident response.

Why is it Important?

Risk Mitigation → Well-defined AI policies minimize legal, financial, and reputational risks caused by unethical AI practices or non-compliance with regulations.

Operational Efficiency → Policies ensure consistency in AI development and streamline decision-making.

Trust and Transparency → Communicating AI policies builds trust with employees and external stakeholders, demonstrating a commitment to responsible AI.

AI Policy Enforcement

Responsibility and Accountability → Clearly state who is responsible for enforcing AI policies (typically the AI Governance Committee or designated team).

Tools and Processes → Employ monitoring tools, regular audits, and established incident response procedures to ensure policies are followed effectively.

Education and Training → All employees impacted by AI policies must understand them and their role in upholding them.

Consequences → There should be clear, proportionate consequences for policy violations to ensure they are taken seriously.

Key Challenges

Keeping pace with regulations → As the AI regulatory landscape evolves, policies must be updated continuously.

Balancing innovation and control → Policies must be strict enough to mitigate risk but flexible enough to allow for responsible innovation.

Enforcement across complex systems → Ensuring adherence to AI policies in distributed systems with multiple AI models can be complex.

Best Practices

Start with clear ethical principles → These form the foundation of your AI policies.

Involve stakeholders → Get input from different departments to ensure policies are practical and effective.

Regularly review and update → AI and regulations change rapidly; your policies must do the same.

CHAPTER 10   AI POLICY MANAGEMENT AND ENFORCEMENT

Benefits of Effective Policy Management → **The advantages of a rigorous approach:**

- Mitigating risks and building trust
- Promoting responsible AI innovation
- Enhancing brand reputation
- Avoiding legal and regulatory penalties
- Achieving long-term sustainability with AI

Here is a comprehensive breakdown of AI policy management and enforcement centered on an AI policy lifecycle management process:

The Imperative for AI Policy Management. Begin by outlining the rapid proliferation of AI systems in various sectors, underscoring the need for clearly defined policies to ensure responsible and ethical AI use. Highlight the potential risks of AI, including

- Bias and discrimination
- Lack of transparency and explainability
- Privacy violations
- Security vulnerabilities
- Unintended consequences and societal harm

Define your AI Policy Management process. Explain AI policy management as the framework of principles, procedures, and standards governing AI development, deployment, and use within your organization. This process should include a continuous lifecycle, ensuring AI aligns with ethical values, legal requirements, and business goals.

CHAPTER 10   AI POLICY MANAGEMENT AND ENFORCEMENT

The AI Policy Lifecycle

Policy Development

- **Collaborative Approach** → Stress the importance of involving diverse stakeholders → data scientists, legal experts, IT, risk management, ethicists, and business domain specialists.

- **Key Policy Areas** → Provide a high-level overview of typical AI policy components.

- **Data Governance** → Data collection, usage, quality, consent, retention, and privacy.

- **Bias and Fairness** → Proactive bias detection, mitigation, and ensuring non-discriminatory outcomes.

- **Explainability and Transparency** → Enabling understanding of AI decisions, especially in high-stakes scenarios.

- **Security and Robustness** → Protecting AI models and data from attacks and ensuring system resilience.

- **Accountability and Liability** → Establishing clear ownership and responsibility for AI outcomes.

- **Ethical Considerations** → Embedding societal values and addressing potential negative impacts.

- **Regulatory Compliance** → Aligning with sector-specific and evolving AI regulations.

- **Tailoring Policies** → Emphasize adapting policies to specific AI use cases and organizational risk tolerance.

CHAPTER 10 AI POLICY MANAGEMENT AND ENFORCEMENT

Policy Dissemination

Provide details on how you plan to make policies easily accessible to all relevant personnel and describe the types of training based on roles and responsibilities. Implement a communication strategy and a multi-faceted communication plan to raise awareness of AI policies and foster a culture of responsible AI throughout the organization.

Policy Implementation

Provide tools, checklists, and templates to guide teams in operationalizing AI policies throughout development. Embed AI policy considerations into the software development lifecycle (SDLC), ensuring checks and reviews at each stage. Establish change management processes for updating AI models and systems, incorporating revisions per evolving policies and regulations.

Policy Monitoring and Enforcement

Define relevant KPIs and metrics for tracking model performance, fairness, explainability and security. Implement proactive auditing schedules to verify compliance with policies and uncover potential issues. Develop protocols for addressing incidents of policy violations, bias, or security breaches and establish a transparent mechanism for addressing policy breaches, ranging from corrective action to retraining or disciplinary measures, depending on severity.

Policy Review and Update

Continuous Improvement Mindset → Highlight the dynamic nature of the AI landscape and the necessity of regularly revisiting policies. Create feedback loop mechanisms for developers, users, and stakeholders to suggest policy improvements.

Monitor Regulatory Landscape → Track emerging AI-specific regulations, ethical guidelines, and industry best practices to evolve your policies proactively.

Understanding the Challenges and Best Practices of Policy Management and Enforcement

CHAPTER 10   AI POLICY MANAGEMENT AND ENFORCEMENT

Acknowledge hurdles like

- Complexity of AI Systems
- Balancing innovation with risk mitigation
- Lack of standardized AI regulations
- Difficulty achieving robust explainability

Best Practices

- **Top-Level Commitment** → Secure executive-level support for AI governance.
- **Centralized Governance Structure** → Consider an AI Governance Center of Excellence.
- **Incentivize Responsible AI** → Reward teams for building ethical and reliable AI systems.
- **Collaborate Externally** → In industry groups, share learnings and influence standards.
- **Proactive Approach** → Foster a mindset of anticipating risks rather than reacting to them.

Conclusion

- **Reiterate the Importance** → Summarize the critical role of AI policy management and enforcement in AI's ethical and sustainable deployment.
- **Call to Action** → Organizations must prioritize AI governance or risk reputational, financial, and legal consequences.

CHAPTER 10   AI POLICY MANAGEMENT AND ENFORCEMENT

Integrating AI Policy Management with Key Business Policies

Cybersecurity → AI policies should address specific cybersecurity risks associated with AI models, such as adversarial attacks and data poisoning. This includes defining security standards for model development and implementation, monitoring, and establishing access controls to AI systems and data.

Risk Management → AI-specific risks, such as bias, lack of explainability, and unintended consequences, must be explicitly incorporated into an organization's risk taxonomy. AI risk assessments should be conducted throughout the model lifecycle, with mitigation plans integrated into broader enterprise risk management strategies.

Compliance → AI policies must reflect regulatory mandates relevant to data privacy (e.g., GDPR, CCPA), sector-specific regulations (e.g., healthcare, finance), and emerging AI-specific legislation. Compliance teams should be involved in AI policy development to ensure alignment with regulations and the organization's risk tolerance.

Incident Management → AI policy should guide incident response plans specifically addressing AI-related failures, security breaches, or ethical breaches. These plans should detail AI teams' reporting mechanisms, roles, responsibilities, and the process for notifying regulatory bodies or affected individuals when required.

Disaster Recovery and Business Continuity Management → AI policies must consider the potential impact of system failures or outages on critical business processes. They must include AI models and data in disaster recovery plans, prioritize critical AI applications for restoration, and ensure continuity plans consider the availability of AI personnel and specialized resources.

CHAPTER 10   AI POLICY MANAGEMENT AND ENFORCEMENT

Privacy → AI policies must prioritize data protection principles in alignment with regulations such as the GDPR and CCPA. They should define clear data collection, storage, usage, and consent guidelines, ensuring transparency about utilizing personal data within AI models. Additionally, policies must address privacy impact assessments for AI projects and processes for handling privacy breaches and upholding individuals' data rights.

Ethics → AI policies should establish a core set of ethical principles that guide the development and deployment of AI systems. These principles should include promoting fairness and preventing discrimination, ensuring transparency and explainability, protecting user privacy, and considering the broader societal impacts of AI. Additionally, policies should include processes for ethical review of AI projects, particularly those with high potential for harm, aiming to proactively address ethical dilemmas and potential negative consequences.

Summary and Thoughts

AI policy management and enforcement must seamlessly integrate with an organization's existing policy and oversight structures to ensure a consistent and holistic approach to governance. This involves aligning AI-specific policies with broader principles around data privacy, cybersecurity, ethics, and risk management. Leveraging established review and enforcement processes and adapting existing compliance monitoring tools creates a robust governance framework for AI. Additionally, by aligning AI governance with overarching compliance reporting and auditing cycles, organizations can ensure a unified view of risk and adherence, reducing redundancies and fostering efficient oversight.

Quiz

1. What is the primary goal of AI policy management?

    A) To limit innovation

    B) To create data silos

    C) To ensure responsible and consistent AI use

    D) To avoid regulatory compliance

2. Which stage of the AI policy lifecycle involves monitoring and ensuring compliance?

    A) Policy drafting

    B) Policy approval

    C) Policy implementation

    D) Policy enforcement

3. Which of the following is NOT a focus area for AI policy management?

    A) Data governance

    B) Model development

    C) Regulatory compliance

    D) Personal preferences

CHAPTER 10 AI POLICY MANAGEMENT AND ENFORCEMENT

4. What should AI policy management integrate with to ensure comprehensive governance?

    A) Key business policies

    B) Personal opinions

    C) Non-relevant guidelines

    D) Outdated standards

5. True or False: AI policy management ensures the consistent and responsible use of AI technologies across an organization.

6. True or False: Policy enforcement is not necessary once policies are drafted and approved.

7. True or False: Integrating AI policy management with key business policies enhances overall governance effectiveness.

8. AI policy management involves creating, implementing, and enforcing policies to ensure the _____ use of AI.

9. The AI policy lifecycle includes drafting, approval, implementation, and _____.

10. Integrating AI policy management with key business policies ensures a _____ approach to governance.

CHAPTER 11

Maintaining Privacy Within Your AI Governance Model

A robust AI governance model prioritizes privacy by embedding privacy by design principles, enforcing strict data governance, and strategically implementing privacy-enhancing technologies to protect sensitive information throughout the AI lifecycle.

Let's outline a strategy for prioritizing privacy within your AI governance model.

Privacy by Design and Default → Embed privacy principles into the foundation of your AI governance framework. Adopt a Privacy by Design approach, where privacy considerations are baked into the earliest stages of AI model development. Set privacy-protective defaults, such as minimizing data collection, utilizing anonymization or pseudonymization techniques where applicable, and respecting user preferences for data sharing.

Strict Data Governance → Implement comprehensive data governance policies specifically addressing AI use cases. This includes clear guidelines on data collection, storage, access controls, consent management, data quality, and retention periods. Enforce strict data lineage tracking to understand the origin and transformations of data used in AI models. Regularly review and update these policies in alignment with evolving data privacy regulations.

Privacy-Enhancing Technologies and Processes → Proactively implement tools and techniques that promote privacy preservation throughout the AI lifecycle. Consider

- **Differential Privacy** → Adding strategic noise to datasets to protect individual privacy while preserving statistical insights.

- **Federated Learning** → Models are trained on decentralized data, minimizing the need to share sensitive information.

- **Homomorphic Encryption** → This technology enables computation on encrypted data without decryption, enhancing privacy in cloud or collaborative AI scenarios.

- **Privacy Impact Assessments** → Mandate these assessments for AI projects, identifying and mitigating potential privacy risks.

Here is a breakdown of privacy throughout the AI lifecycle:

Development Phase

Integrate privacy as a core design principle in your AI systems from the outset. Minimize collecting personally identifiable information (PII) and prioritize using anonymized or synthetic data whenever possible.

Data Minimization. Collect only the data strictly necessary for the AI model's intended purpose. Implement strict access controls and secure storage mechanisms to protect sensitive data during development.

Validation Phase

Consider using differential privacy techniques when validating AI models on sensitive datasets. This involves strategically adding noise to the data to obscure individual details while preserving overall statistical trends.

Ensure your validation datasets are diverse and representative to mitigate potential biases that might disproportionately impact certain groups based on protected characteristics.

Deployment Phase

Obtain clear and informed consent from individuals before using their data in AI models. Provide user-friendly transparency about how their data is used and allow them to manage their privacy preferences.

Strong encryption protects AI models and associated data in production environments at rest and in transit. Implement strict access controls and role-based permissions to limit who can access sensitive information.

Monitoring Phase

Schedule regular audits to assess privacy compliance throughout the AI model's lifecycle. Monitor for potential data leakage, unintended re-identification of individuals, or discriminatory outputs.

Incident Response. Establish clear protocols for responding to privacy breaches. This includes promptly identifying the source, notifying affected individuals, containing the breach, and, if necessary, reporting it to regulatory authorities.

As regulations evolve and new privacy-preserving technologies emerge, continuously update your AI models and processes to ensure ongoing protection of user data.

Maintaining privacy in AI is an ongoing process. By embedding privacy considerations at every stage of the lifecycle, promoting transparency, and remaining vigilant with monitoring, you can build trust in your AI systems while responsibly leveraging their benefits.

CHAPTER 11   MAINTAINING PRIVACY WITHIN YOUR AI GOVERNANCE MODEL

AI Actors Collaborating with Legal Counsel on AI Privacy Matters

In the era of rapid AI advancement, safeguarding privacy has become a paramount concern for organizations across all sectors. In close collaboration with legal counsel, a scalable AI Centre of Excellence (CoE) plays a crucial role in proactively addressing privacy challenges. This section explores the vital partnership between AI experts and legal professionals in developing privacy-centric AI policies, conducting impact assessments, implementing privacy-enhancing technologies, providing essential training, establishing monitoring mechanisms, and navigating the ever-evolving regulatory landscape. By prioritizing privacy throughout the AI lifecycle, organizations can foster trust, mitigate risks, and ensure this transformative technology's responsible and ethical use.

AI Actors and Legal Counsel Collaboration

Here is a breakdown of how AI resources and legal counsel can collaborate to prioritize privacy in an AI Center of Excellence (CoE) and oversight programs.

    Developing Comprehensive AI Privacy Policies → The AI CoE should work closely with legal counsel to craft detailed privacy policies tailored to the organization's specific AI use cases. These policies should align with relevant data privacy regulations (GDPR, CCPA, etc.), industry-specific standards, and the organization's risk appetite. They should address issues like data collection, use limitations, consent, retention, third-party data sharing, and individuals' rights (access, rectification, erasure).

    Conducting Privacy Impact Assessments (PIAs) → Mandate PIAs for all high-risk AI projects (especially those with sensitive data). Legal counsel should guide the PIA process, ensuring it thoroughly evaluates potential privacy risks, proposes mitigation measures, and documents compliance efforts. AI experts within the CoE should collaborate by providing technical insights into data use and potential vulnerabilities.

Designing Privacy-Enhancing Technologies (PETs) → The AI CoE should actively research and implement PETs. Legal counsel can help evaluate the compliance implications and effectiveness of techniques like differential privacy, federated learning, and homomorphic encryption. Collaboration in this space ensures that privacy is not an afterthought but a driving force in AI innovation.

Training and Awareness → Legal counsel should play a key role in developing privacy training programs for the AI CoE, data scientists, and project managers. This training should cover regulatory requirements, ethical principles, data protection best practices, and the organization's privacy policies. Regular awareness campaigns across the company reinforce the importance of privacy throughout the AI lifecycle.

Monitoring and Incident Response → The AI CoE and legal team should collaborate on establishing processes for monitoring data privacy compliance within AI systems. This includes regular audits, metrics to track data usage and mechanisms for detecting potential privacy violations. Together, they must develop a robust incident response plan that outlines reporting, investigation, remediation, and regulatory notification procedures in case of a privacy breach.

Staying Updated on Evolving Regulations → The AI and legal teams must form a proactive partnership to stay informed about the rapidly evolving data privacy landscape. They should monitor new and proposed legislation, analyze the implications for the organization's AI initiatives, and ensure policies and technology remain compliant by participating in industry working groups and conferences, which aids in staying ahead of the curve.

CHAPTER 11   MAINTAINING PRIVACY WITHIN YOUR AI GOVERNANCE MODEL

Summary and Thoughts

AI teams, including data scientists, engineers, and AI project managers, provide critical technical expertise on data collection, model development, and the potential for privacy risks within AI systems. They must collaborate with legal counsel to design and implement AI solutions that align with technical requirements and privacy regulations. Legal counsel will guide these teams on interpreting regulations like the GDPR and CCPA, ensuring that AI systems comply with data minimization principles, biometrics, consent, transparency, and individual rights.

AI actors and legal counsel must work proactively to conduct privacy impact assessments, identify potential risks, and develop mitigation strategies. They design privacy-enhancing technologies into AI systems from the outset and create protocols for monitoring, auditing, and responding to potential privacy breaches. This collaborative approach ensures that privacy is not an afterthought but a foundational pillar of responsible AI development, enabling organizations to leverage AI's transformative power while respecting individual privacy rights.

Quiz

1. What is a key consideration when maintaining privacy in AI governance?

    A) Ignoring data protection laws

    B) Minimizing data usage

    C) Ensuring compliance with privacy regulations

    D) Avoiding data anonymization

CHAPTER 11   MAINTAINING PRIVACY WITHIN YOUR AI GOVERNANCE MODEL

2. Who should AI actors collaborate with to address privacy matters?

   A) Marketing team

   B) Legal counsel

   C) Sales department

   D) Finance team

3. Which of the following is NOT a privacy risk in AI governance?

   A) Data breaches

   B) Unauthorized data access

   C) Transparent data usage

   D) Data anonymization

4. What is essential for AI systems to handle sensitive data responsibly?

   A) Avoiding privacy regulations

   B) Implementing strong data governance practices

   C) Using data without consent

   D) Ignoring data protection measures

5. True or False: Collaborating with legal counsel helps AI actors address privacy risks effectively.

6. True or False: Ensuring compliance with privacy regulations is optional for AI governance.

7. True or False: Maintaining privacy in AI governance involves implementing strong data governance practices.

CHAPTER 11   MAINTAINING PRIVACY WITHIN YOUR AI GOVERNANCE MODEL

8. AI actors should collaborate with _____ to address privacy matters and ensure compliance with regulations.

9. Maintaining privacy within AI governance involves addressing risks such as data breaches, unauthorized data access, and lack of _____.

10. Strong data governance practices are essential for AI systems to handle _____ data responsibly.

CHAPTER 12

Human Oversight of AI Systems

Meaningful human oversight is a fundamental pillar of responsible AI governance, essential for preventing unintended consequences, upholding accountability, and building trust in AI systems. Organizations can mitigate AI's inherent limitations by ensuring a calibrated blend of human expertise and AI capabilities, promoting ethical decision-making, and harnessing the technology's transformative potential safely and responsibly. Ensuring meaningful human oversight in AI decision-making processes to prevent unintended consequences and enable human intervention when needed is a key component of an AI governance program.

The Inherent Limitations of AI

While AI systems excel at pattern recognition and complex calculations, they lack the nuanced understanding, critical thinking, and ethical judgment humans possess. AI models can inherit biases from their training data, produce flawed results due to errors or overfitting, and struggle to adapt to unanticipated scenarios outside their training domain.

CHAPTER 12   HUMAN OVERSIGHT OF AI SYSTEMS

Preventing Unintended Consequences

Unfettered AI decision-making can have unintended consequences, particularly in high-stakes domains like healthcare, finance, or criminal justice. These consequences might include discriminatory outcomes, safety hazards, reputational damage, or financial losses. Meaningful human oversight acts as a safeguard, allowing humans to identify potential errors, biases, or ethical concerns before they cause harm.

The Importance of Human Judgment

In many cases, AI should be a decision-support tool rather than an autonomous decision-maker. Human judgment is essential for interpreting AI outputs in context, considering broader implications beyond the model's scope, and applying ethical principles that might not be fully encoded in the AI system.

Enabling Human Intervention

AI governance programs must establish clear mechanisms for human intervention when necessary. This includes defining escalation paths, identifying roles and responsibilities for overriding AI decisions, and documenting justifications for human intervention. A well-defined process ensures timely and appropriate human intervention to mitigate risks and uphold organizational values.

Transparency and Explainability

For human oversight to be effective, AI systems must be designed with explainability in mind. This means understanding how a model reached a particular decision or recommendation. Explainability empowers human experts to scrutinize AI outputs, detect potential flaws, and intervene, when necessary, with informed judgment.

Accountability and Responsibility

Meaningful human oversight reinforces accountability within AI systems. Ensuring that humans remain involved in the decision-making loop makes it clear where responsibility lies for the outcomes of those decisions. This helps prevent the diffusion of responsibility when relying solely on opaque, automated systems.

Building Trust
Public and stakeholder trust in AI is essential for its widespread adoption. Demonstrating a commitment to human oversight signals that an organization is taking responsibility for the ethical and safe use of AI. This transparency helps build trust and mitigate the fear that AI will replace humans in critical roles.

Balancing Innovation and Risk
Human oversight should not be seen as a barrier to innovation. Instead, it should be a framework for responsible innovation. Organizations can confidently explore new AI applications while minimizing unintended consequences by carefully considering potential risks and ensuring human involvement in critical decisions.

Designing for Human-AI Collaboration
The goal should be to leverage the strengths of both humans and AI. Design AI systems and workflows that facilitate collaboration, where humans can provide feedback, refine models, and override decisions, enabling a continuous learning loop that improves the overall system.

AI Governance Best Practice
Ensuring meaningful human oversight is a fundamental component of robust AI governance programs. Organizations must define clear policies, implement technical solutions for interpretability, and establish training and review processes to support effective human oversight. By prioritizing human involvement in the decision-making process, organizations can unlock the benefits of AI while safeguarding against potential pitfalls.

Human Oversight Challenges and Risk Mitigation

Meaningful human oversight is a fundamental pillar of responsible AI governance, essential for preventing unintended consequences, upholding accountability, and building trust in AI systems. Organizations can mitigate AI's inherent limitations by ensuring a calibrated blend of human expertise and AI capabilities, promoting ethical decision-making, and harnessing the technology's transformative potential safely and responsibly. Ensuring meaningful human oversight in AI decision-making processes to prevent unintended consequences and enable human intervention when needed is a key component of an AI governance program.

The Inherent Limitations of AI

While AI systems excel at pattern recognition and complex calculations, they lack the nuanced understanding, critical thinking, and ethical judgment humans possess. AI models can inherit biases from their training data, produce flawed results due to errors or overfitting, and struggle to adapt to unanticipated scenarios outside their training domain.

Preventing Unintended Consequences
Unfettered AI decision-making can have unintended consequences, particularly in high-stakes domains like healthcare, finance, or criminal justice. These consequences might include discriminatory outcomes, safety hazards, reputational damage, or financial losses. Meaningful human oversight acts as a safeguard, allowing humans to identify potential errors, biases, or ethical concerns before they cause harm.

The Importance of Human Judgment
In many cases, AI should be a decision-support tool rather than an autonomous decision-maker. Human judgment is essential for interpreting AI outputs in context, considering broader implications beyond the model's scope, and applying ethical principles that might not be fully encoded in the AI system.

Enabling Human Intervention

AI governance programs must establish clear mechanisms for human intervention when necessary. This includes defining escalation paths, identifying roles and responsibilities for overriding AI decisions, and documenting justifications for human intervention. A well-defined process ensures timely and appropriate human intervention to mitigate risks and uphold organizational values.

Transparency and Explainability

For human oversight to be effective, AI systems must be designed with explainability in mind. This means understanding how a model reached a particular decision or recommendation. Explainability empowers human experts to scrutinize AI outputs, detect potential flaws, and intervene, when necessary, with informed judgment.

Accountability and Responsibility

Meaningful human oversight reinforces accountability within AI systems. Ensuring that humans remain involved in the decision-making loop makes it clear where responsibility lies for the outcomes of those decisions. This helps prevent the diffusion of responsibility when relying solely on opaque, automated systems.

Building Trust

Public and stakeholder trust in AI is essential for its widespread adoption. Demonstrating a commitment to human oversight signals that an organization is taking responsibility for the ethical and safe use of AI. This transparency helps build trust and mitigate the fear that AI will replace humans in critical roles.

Balancing Innovation and Risk

Human oversight should not be seen as a barrier to innovation. Instead, it should be a framework for responsible innovation. Organizations can confidently explore new AI applications while minimizing unintended consequences by carefully considering potential risks and ensuring human involvement in critical decisions.

Designing for Human-AI Collaboration
The goal should be to leverage the strengths of both humans and AI. Design AI systems and workflows that facilitate collaboration, where humans can provide feedback, refine models, and override decisions, enabling a continuous learning loop that improves the overall system.

AI Governance Best Practice
Ensuring meaningful human oversight is a fundamental component of robust AI governance programs. Organizations must define clear policies, implement technical solutions for interpretability, and establish training and review processes to support effective human oversight. By prioritizing human involvement in the decision-making process, organizations can unlock the benefits of AI while safeguarding against potential pitfalls.

Human Oversight Challenges and Risk Mitigation

Let's delve into the challenges of human oversight in AI decision-making and outline potential mitigation strategies.

Challenge 1 → Cognitive Overreliance
When humans are presented with AI-generated recommendations, there is a risk of overreliance. Humans might assume the AI system is more accurate or objective than it is, leading to reduced scrutiny and a failure to identify potential errors or biases.
Mitigation → Train personnel on AI's limitations, encourage critical thinking, and emphasize that AI outputs should be carefully evaluated rather than blindly accepted.

Challenge 2 → Information Overload and Complexity
Understanding the logic behind a decision can be overwhelming in complex AI systems, even with explainability tools. Humans might struggle to grasp the intricate relationships between variables, making it difficult to provide meaningful oversight.

Mitigation → Focus on explainability techniques tailored to the human reviewer's expertise. Provide clear visualizations, simplify explanations, and offer training on interpreting AI outputs.

Challenge 3 → Time and Resource Constraints

A thorough review of AI decisions can be time-consuming, especially in high-volume or real-time scenarios. Organizations might lack the human resources or expertise to provide effective oversight consistently.

Mitigation → Implement a risk-based approach. Prioritize human oversight for high-risk decisions. Utilize automated tools to flag potentially problematic cases for more in-depth human review.

Challenge 4 → Lack of Human Expertise

Human experts might not understand AI models enough to provide a meaningful assessment in highly specialized domains. This knowledge gap can hinder effective oversight and evaluation.

Mitigation → Invest in cross-training programs. Educate domain experts about AI concepts and train data scientists in domain-specific knowledge to foster better collaboration.

Challenge 5 → Alert Fatigue

If AI systems generate too many alerts or frequently flag potential issues, humans might become desensitized, which could lead to missed critical problems.

Mitigation → Carefully calibrate alert thresholds to avoid false positives. Provide ongoing training on the importance of each alert and the potential consequences of inaction.

Challenge 6 → Complacency and Misplaced Trust

Over time, as AI systems demonstrate reliability, humans might become complacent and less vigilant in their oversight role.

Mitigation → Regularly audit AI systems and decisions, even when performance appears satisfactory. Reinforce the importance of ongoing oversight through training and awareness campaigns.

Challenge 7 → Misaligned Incentives
If humans are incentivized to prioritize speed or volume over careful review, they might be less likely to intervene in AI decisions, even when necessary.
Mitigation → Align performance metrics with responsible AI principles, rewarding thoroughness, ethical considerations, and efficiency.

Challenge 8 → Difficulty Justifying Overrides
Humans might hesitate to override AI decisions, fearing scrutiny or being unable to justify their reasoning, particularly when AI performs fully.
Mitigation→ Create a clear process for documenting the rationale behind overrides and establish a culture where human judgment is valued and supported.

Challenge 9 → Limited Scalability
Human oversight can be a bottleneck, especially as AI deployment increases. Scaling human oversight effectively while maintaining rigor is an ongoing challenge.
Mitigation → Explore hybrid approaches. Utilize AI tools to assist humans in prioritizing cases for review while reserving in-depth analysis for the most critical decisions.

Challenge 10 → Evolving Technology
AI is rapidly advancing, and new techniques like deep learning pose increasing challenges for explainability. To keep pace, oversight strategies must be adaptable.
Mitigation → Invest in research and development on explainable AI (XAI). Engage with the research community to stay updated on the possibilities and limitations of explainability methods.

Summary and Thoughts

Human oversight in AI decision-making offers significant benefits. It helps prevent unintended consequences, ensures accountability, builds trust, and allows humans to apply ethical judgment and contextual

CHAPTER 12   HUMAN OVERSIGHT OF AI SYSTEMS

understanding that AI systems might lack. By enabling timely intervention when necessary, human oversight serves as a crucial safeguard in the responsible deployment of AI.

However, implementing effective human oversight poses several challenges. These include the potential for overreliance on AI, complexity of models hindering understanding, time constraints, lack of expertise, alert fatigue, and difficulty scaling as AI use expands. Organizations must be aware of these challenges and proactively develop mitigation strategies. This involves investing in explainability, risk-based review prioritization, continuous training, and fostering a culture where responsible AI and human judgment are valued.

Quiz

1. What is a key challenge in human oversight of AI systems?

    A) Lack of technology

    B) Ensuring human involvement at critical decision points

    C) Ignoring ethical considerations

    D) Avoiding accountability

2. Which limitation is inherent to AI systems?

    A) Perfect accuracy

    B) Lack of bias

    C) Inability to understand context like humans

    D) Complete transparency

CHAPTER 12   HUMAN OVERSIGHT OF AI SYSTEMS

3. What is essential to mitigate risks associated with AI systems?

   A) Ignoring potential biases

   B) Comprehensive human oversight

   C) Avoiding transparency

   D) Disregarding ethical concerns

4. Which of the following is NOT a challenge in human oversight of AI systems?

   A) Ensuring accountability

   B) Lack of human involvement

   C) Monitoring AI decisions

   D) Over-reliance on AI without checks

5. True or False: Human oversight is crucial for mitigating risks associated with AI systems.

6. True or False: AI systems can fully understand context and make decisions without human input.

7. True or False: Ensuring human involvement at critical decision points helps address ethical concerns in AI governance.

8. Comprehensive _____ is essential to mitigate risks associated with AI systems.

9. AI systems have inherent limitations, such as the inability to understand _____ like humans do.

10. Human oversight involves ensuring accountability and monitoring AI decisions to address _____ concerns.

CHAPTER 13

The Power of Stakeholder Engagement in AI Governance

Strategies for Inclusive and Effective Decision-Making

To keep pace with AI's rapid evolution, immense potential, and significant risks, a collaborative and inclusive approach to governance is essential to realize the full benefits of AI. Stakeholder engagement, involving diverse perspectives across sectors, is crucial for ensuring AI development aligns with societal values, builds trust, and avoids unintended consequences. This chapter provides insights and strategies for effectively managing stakeholder engagement throughout the AI governance process.

CHAPTER 13 THE POWER OF STAKEHOLDER ENGAGEMENT IN AI GOVERNANCE

Why Stakeholder Engagement Matters

Ethical AI Development → Stakeholders from civil society, academia, and marginalized groups can raise ethical concerns, potential biases, and unintended consequences, ensuring AI systems are developed and deployed responsibly.

Enhanced Public Trust → Transparency and inclusivity in governance foster trust. Involving the public in decision-making can dispel fears and promote greater acceptance of AI technologies.

Regulatory Compliance → Understanding diverse stakeholder needs helps create frameworks that support innovation while protecting individuals and addressing concerns like privacy and fairness.

Improved Innovation → Stakeholder insights can uncover new applications for AI, address unmet needs, and ensure technologies benefit society as a whole.

Mitigating Risks → Engaging a broad range of stakeholders early on helps identify potential negative impacts of AI, allowing developers and policymakers to address them proactively.

Identifying Key Stakeholders

Internal Stakeholders:

- Data scientists and AI developers
- Business units utilizing AI applications
- Legal, compliance, and risk management teams
- Company leadership and board of directors

External Stakeholders:

- Customers and end users
- Industry partners and competitors
- Regulatory bodies
- Civil society organizations, advocacy groups
- Academia (AI researchers, ethicists, social scientists)
- The public and the communities your organization serves

CHAPTER 13   THE POWER OF STAKEHOLDER ENGAGEMENT IN AI GOVERNANCE

Mapping Stakeholders. Create a stakeholder map, like Figure 13-1 below, to visualize AI initiatives' potential impact and influence. Prioritize those with elevated or multiple levels of responsibilities for increased and in-depth engagement. *Many of the role descriptions in the stakeholder map are described in Chapter 7.*

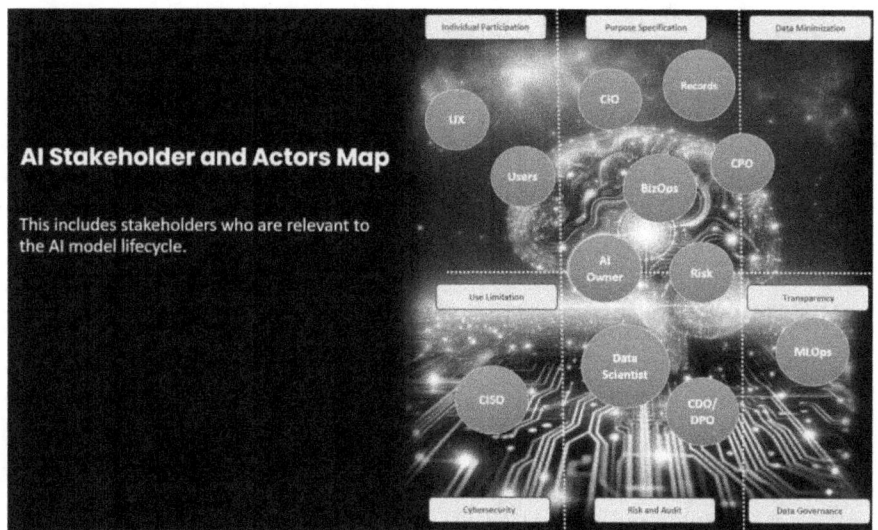

Figure 13-1. *AI Stakeholder Map*

Strategies for Meaningful Engagement

Start Early and Iterate → Involve stakeholders from the initial stages of AI project planning and continue engagement throughout the project's lifecycle.

Tailor Communication → Adjust messaging and engagement channels to suit different stakeholder groups. Avoid overly technical jargon and be transparent about AI's potential and limitations.

Diverse Engagement Methods

- Public forums and consultations
- Workshops and focus groups

- Collaborative partnerships with academia or civil society
- Online surveys or feedback platforms
- Advisory boards with representation from various groups

Establish a Clear Purpose → Define the objective of stakeholder engagement (e.g., input on AI ethics guidelines and feedback on a new model) to foster focused participation.

Manage Expectations → Be transparent about the level of influence stakeholders will have over decisions and the limitations of your AI governance process.

Overcoming Challenges

Conflicting Interests → Acknowledge different priorities and seek common ground. Encourage compromise and find ways to highlight areas where goals align.

Limited Resources → Smaller organizations may need creative solutions. Leverage online platforms, partner with universities, or seek pro bono support from civil society organizations.

Ethical Considerations Take Center Stage → Stakeholders from academia and civil society organizations often champion ethical considerations. Their involvement ensures ethical principles are embedded within AI governance frameworks, mitigating bias and unintended consequences.

Building Public Trust and Confidence → The public may have anxieties or a limited understanding of AI. Invest in public education and simplify complex topics for better engagement. Engaging citizens through workshops and educational forums helps demystify AI and empowers individuals to participate effectively in future discussions about its role in society. Transparency and inclusivity are key to building public trust in AI. Engaging citizens in discussions about AI applications fosters greater understanding and acceptance.

Robust Regulatory Frameworks → Regulatory bodies need a comprehensive understanding of AI's impact across different sectors. Collaborative dialogues with stakeholders inform the development of regulations that encourage innovation and protect public interests.

Uncovering Unforeseen Challenges → From potential job displacement to privacy concerns, AI raises a spectrum of issues. Stakeholders across sectors bring unique experiences, helping identify and address unforeseen challenges before they escalate.

A Broader Range of Applications → Stakeholders with diverse needs and expertise can identify innovative AI applications that benefit society. For example, collaboration with healthcare professionals might lead to AI-powered diagnostic tools, while environmental groups could contribute ideas for sustainable AI solutions.

Strengthened Collaboration and Innovation → When stakeholders from academia, industry, and civil society collaborate, innovation thrives. Joint research initiatives and knowledge-sharing platforms foster creative solutions to complex challenges, maximizing AI's potential for good.

Mitigating Risks Proactively → Open communication across stakeholder groups allows concerns to be raised early on. Identifying potential risks early in the development cycle allows for more proactive mitigation strategies, avoiding costly and damaging consequences down the line.

Building a Future for All → AI has the potential to revolutionize how we live and work. Involving a broad range of stakeholders in AI governance ensures that this technology serves all needs, not just a select few. We can pave the way for a responsible and equitable AI future by fostering inclusivity, transparent communication, and collaboration. Finding the right voices can help avoid tokenism. Contact diverse communities to ensure genuine representation, not just the loudest or most privileged voices.

CHAPTER 13   THE POWER OF STAKEHOLDER ENGAGEMENT IN AI GOVERNANCE

Building Public Trust and Involvement in AI

AI governance thrives when it includes diverse voices. Stakeholder engagement enhances ethical concerns, builds public trust, designs better regulations, identifies hidden risks, and ensures AI benefits everyone. This is possible because diverse groups (industry experts, ethicists, and the public) bring unique concerns and insights. This multi-faceted input leads to more robust AI systems that avoid unintended consequences.

Ordinary people might feel unsure of how to get involved. Staying informed matters. Read reports from AI think tanks and news articles on current AI applications and keep track of any proposed new regulations. Participate when companies or government agencies request public input on a specific AI project! Share your voice through surveys, online forums, or town hall meetings.

Social media is a tool for involvement. Share articles, join online discussions about AI ethics, and voice your thoughts respectfully. Civil society organizations focused on AI ethics often need support and amplification. Donating, volunteering, or simply sharing their calls to action can make a difference. It is never too late to reach out to your elected officials to express your views on how AI should be developed and regulated in a way that impacts your community. Even seemingly small actions, when combined with others doing the same, ensure AI is a force for good that works for the benefit of society.

Stakeholder engagement is not a one-time task but an ongoing commitment. Effective AI governance is adaptable, incorporating lessons learned through stakeholder feedback. An inclusive approach is vital for building trust, responsible innovation, and an AI future that benefits all.

Breakdown of Common AI Myths and Strategies to Debunk Them

Myth 1 → AI Will Achieve Superintelligence and Take Over

This sci-fi-fueled fear is persistent. Today's AI is highly specialized. While impressive, AI systems lack the general intelligence or self-preservation instincts to pose an existential threat. Emphasizing AI's limitations compared to human cognition can dispel this misconception.

Myth 2 → AI Will Cause Mass Unemployment

AI will automate some tasks, and jobs will evolve. History shows emerging technologies tend to create new employment alongside displacement. The focus should be on reskilling initiatives, education aligning with shifting needs, and policies addressing equitable distribution of AI-generated wealth.

Myth 3 → AI is Unbiased and Objective

AI models learn from data, and that data often reflects existing biases in society. Unchecked, this can lead to discriminatory outcomes. Promoting transparency about training data, active bias detection, and diverse development teams are essential to combatting this perception and building fairer AI.

Myth 4 → AI is Too Complex for the Average Person to Understand

This creates a feeling of powerlessness. Efforts to break down AI into simpler terms using analogies and explaining basic concepts demystify the technology. Showcasing real-world examples of AI people already use (like recommendation engines) makes it less intimidating.

Myth 5 → AI Is Only for Big Tech Companies

While large corporations have resources, AI is increasingly accessible. Smaller businesses can leverage pre-trained models, cloud-based AI services, and open-source tools. Highlighting these options and offering resources for small business adoption changes this narrative.

Addressing AI Myths

A proactive approach to changing myths about AI centers on public education. This entails clear communication from AI creators, explaining their systems' work and acknowledging potential benefits and limitations. Accessible workshops, engaging online resources, and media portrayals focusing on balanced depictions of AI are crucial to combat sensationalized narratives. Educating the public empowers individuals to differentiate between realistic capabilities and far-fetched scenarios.

Beyond education, a collaborative effort is needed. Policymakers, industry leaders, and educators must work in tandem. This involves crafting responsible AI guidelines, promoting ethical development practices, and addressing AI's potential societal impacts. Proactively emphasizing transparency, accountability, and human-centered design can foster a better-informed public perception of AI. This ultimately paves the way for trust and adoption of AI technologies that benefit society.

While public education and collaboration are vital, changing AI myths also demands a fundamental shift in how AI is developed and presented. Companies and researchers overhype AI capabilities, fueling unrealistic expectations and fearmongering. Instead of focusing solely on changing public perception, we must prioritize responsible AI. This means clearly emphasizing limitations alongside potential and actively demystifying "black box" systems through explainability initiatives.

This perspective emphasizes integrity within the AI community itself. Sensationalized marketing must be replaced with realistic presentations of what AI can (and cannot) do today. Proactively inviting public scrutiny, for example, through open-source models or participatory auditing, fosters trust. By placing accountability on AI creators, we create an ecosystem where public perception is shaped by tangible actions, not just educational campaigns.

CHAPTER 13   THE POWER OF STAKEHOLDER ENGAGEMENT IN AI GOVERNANCE

Addressing the AI Skills Gap and Diversity Crisis

During a panel discussion I participated in addressing the AI skills gap and diversity crisis in America, the moderator asked if minority-based businesses have equal access to funding for an AI startup and if universities, specifically historically Black colleges and universities (HBCUs), are teaching AI subject matter. My response was a resounding no, and no. There are barriers when it comes to funding and education for minorities and HBCUs, respectively. Several barriers contribute to the shortage of AI students at HBCUs. Limited funding for cutting-edge computing infrastructure and AI faculty hinders the ability to provide competitive programs. A lack of dedicated AI courses or degree programs creates a less visible pathway for students interested in the field. HBCUs have historically faced chronic underfunding compared to predominantly Caucasian universities. This systemic disadvantage translates into outdated facilities, limited resources, and difficulty attracting and retaining top faculty in fields like AI, where talent competition is high. Many students at HBCUs may lack awareness about the career opportunities and potential impact of AI. Outreach initiatives and partnerships with tech companies are needed to showcase the potential of AI careers and inspire students to pursue this field.

Representation matters. The demand for AI professionals far outstrips supply, creating a skills gap. Furthermore, a diversity crisis looms over the AI field, with minorities, including Black individuals, vastly underrepresented. This lack of diversity risks perpetuating biases and inequality within AI systems. The lack of visible Black AI professionals, researchers, and entrepreneurs creates a perception that the field is inaccessible. Spotlighting HBCU alums who have forged successful AI careers can provide essential role models and mentorship. Targeted funding to enhance infrastructure, establish dedicated AI programs, and support faculty at HBCUs is crucial. Scholarships and internships specifically designed for HBCU students can create pathways into AI careers. Collaborations between HBCUs and tech companies have the potential to offer promising avenues. These can provide mentorship, resources, research opportunities, and real-world projects, bridging the gap between classroom learning and industry application. Introducing AI concepts early through K-12 outreach programs within communities served by HBCUs can ignite interest and build foundational skills. These programs are essential to creating a pipeline of diverse AI talent.

Addressing the shortage of AI students at HBCUs requires a multi-faceted effort. Government, industry, educational institutions, and community organizations must play a role in investing resources, creating opportunities, and dismantling long-standing barriers to ensure HBCUs become leaders in preparing the next generation of diverse AI leaders. HBCUs have a long history of producing exceptional Black talent across various disciplines. As AI revolutionizes industries, these institutions have the potential to become powerhouses of AI education, nurturing a diverse workforce to shape the future of this transformative technology.

Systemic Barriers to Minority AI Entrepreneurs

Despite the enormous potential of artificial intelligence to drive innovation, the field suffers from a stark lack of diversity. Minority-owned AI startups are significantly underrepresented, reflecting systemic barriers that impede their ability to launch and thrive. Several factors fuel this inequality, including a lack of access to early-stage capital, limited networks within the venture capital ecosystem, and unconscious biases that can hinder fair evaluation of their ideas. Securing funding is a universal challenge for startups, but minority-owned AI ventures often face an even steeper uphill battle. Traditional venture capital firms tend to favor founders who fit a specific mold, usually mirroring the demographics of those leading the firms themselves. This can result in minority entrepreneurs being overlooked or undervalued due to unconscious biases or a lack of familiarity with their unique challenges. Furthermore, smaller minority-owned firms might not have extensive networks or established track records, which investors frequently rely upon to mitigate risk.

The lack of funding for minority-owned AI startups has far-reaching consequences. It stifles innovation, as potentially groundbreaking ideas may never receive the support needed to reach fruition. Additionally, it perpetuates the cycle of wealth inequality, further exacerbating the economic disparities within society. Without diverse voices and perspectives shaping AI development, we risk creating technologies that perpetuate biases and fail to address the needs of underrepresented communities.

Overcoming this funding gap requires a multifaceted approach. Increased awareness and education among investors about unconscious biases are crucial. Programs designed to provide mentorship, networking opportunities, and access to seed funding for minority entrepreneurs can make a significant difference. Additionally, intentional efforts by major venture capital firms to diversify their investment teams and actively seek out minority-led startups will foster a more equitable and inclusive AI landscape.

Creating funding programs, incubators, and accelerators specifically designed for minority entrepreneurs in the AI field can provide significant support. These initiatives can offer tailored mentorship, networking opportunities with relevant investors, and access to seed funding during a startup's crucial preliminary stages. Additionally, building strong networks and communities for minority AI founders fosters collaboration, knowledge sharing, and the chance to connect with potential investors who value diversity and inclusion.

Combating unconscious bias among investors is crucial. This involves educational programs and workshops to raise awareness within venture capital firms and encourage diversity in hiring practices. Actively seeking out and including minority investors in decision-making can lead to fairer evaluation processes and a deeper understanding of minority-led AI startups' unique potential and challenges. By increasing representation in the investment landscape, we normalize seeing minority founders and entrepreneurs in these spaces, slowly breaking down the subconscious barriers that often hinder funding decisions.

Tailoring the Approach of Changing AI Myths for Diverse Cultural and Racial Groups

Different communities may have experienced the negative impacts of technology-driven bias and discrimination. Acknowledging these experiences and how they shape perceptions of AI is crucial. Be mindful of the history of technology being developed by privileged groups and deployed in ways that may disadvantage others. This requires approaching conversations with humility and a willingness to listen.

Culturally Relevant Messaging

Use relatable examples and analogies drawn from the cultural context of your target audience. For example, a farming community might better understand AI concepts framed in terms of crop optimization rather than abstract ideas unrelated to their lived experiences.

Provide resources in multiple languages, not just the dominant one. Partnering with community translators ensures accurate messaging, not just literal translation, to nuancedly convey potential benefits and risks.

Involve community leaders, elders, or culturally respected figures in AI education efforts. Their endorsement and participation build trust.

Address Specific Concerns

Different communities may fear unique things about AI. For example, some groups might worry about surveillance, while others are anxious about AI replacing jobs vital to their local economy. Engage in open dialogue to address these concerns directly.

Emphasize that AI can perpetuate existing biases, but it can also be a tool to expose and fight those biases. Give examples of AI used to address inequality in healthcare or access to services.

Empowerment, Not Just Education

Go beyond explaining AI to equip people with basic AI skills. This could mean coding workshops, data literacy programs, or tools that allow citizens to examine datasets for potential bias.

Invite diverse communities to participate in developing AI solutions that meet their needs. This fosters ownership and counters the idea of AI being imposed from the outside.

Highlight role models and showcase individuals from diverse backgrounds who are successful in AI development, research, or ethics. This challenges stereotypes and inspires.

Changing AI myths across cultures and races requires a sensitive and adaptable approach. Acknowledge the historical context surrounding technology and its potential for bias, and work to establish trust by using culturally relevant examples and involving respected community figures. Tailor your message to address specific concerns about job loss or surveillance within distinct groups. Focus on empowerment, not just education, by offering opportunities to build AI literacy and participate in developing AI solutions that directly serve diverse communities. Highlighting role models from underrepresented backgrounds can further inspire and shift perceptions of who creates and benefits from AI technology.

CHAPTER 13 THE POWER OF STAKEHOLDER ENGAGEMENT IN AI GOVERNANCE

Summary and Thoughts

AI governance benefits from diverse stakeholder participation. Businesses, academics, civil society, and the public bring valuable perspectives, promoting ethical development, public trust, better regulations, and broader innovation. It then encourages ordinary people to get involved by staying informed, participating in public consultations, engaging on social media, supporting AI ethics organizations, and contacting elected officials.

My thoughts on diverse stakeholder involvement

- **Have a Strong Message** → The focus on inclusivity is essential for ensuring AI benefits all of society, not just the technically inclined.

- **Put Forth Concrete Action Steps** → The guidance for ordinary people is commendable, moving beyond abstract principles and offering tangible ways to make their voices heard.

- **Make Room for Expansion** → While comprehensive, it could be even stronger by mentioning

  - The importance of finding credible information sources amidst potential AI hype

  - Addressing barriers to participation some groups face (lack of Internet access, limited awareness of opportunities)

HBCUs often face a significant disadvantage in establishing robust AI programs. Underfunding translates into limited resources for innovative computing infrastructure, difficulty attracting AI faculty within a competitive market, and a lack of specialized AI courses or degree programs. These factors hinder HBCUs from fully participating in the AI revolution and cultivating diverse leaders in the field. This shortage

CHAPTER 13   THE POWER OF STAKEHOLDER ENGAGEMENT IN AI GOVERNANCE

of AI programs has far-reaching consequences. It contributes to the underrepresentation of minorities in the AI industry and perpetuates the potential for bias in AI systems. Furthermore, it denies HBCU students access to lucrative career opportunities and the power to shape the future trajectory of this transformative technology.

Systemic barriers faced by minority AI entrepreneurs are multifaceted and deeply ingrained, hindering their success at every stage of their journey. These barriers include limited access to early-stage capital, as investors often favor founders fitting a stereotypical mold. This lack of initial funding makes it harder to build prototypes, attract talent, and secure the credibility needed for larger investments. Additionally, a lack of strong networks within established tech and venture capital circles limits minority entrepreneurs' access to mentorship, crucial connections, and knowledge of funding channels. Unconscious bias among investors and gatekeepers can further result in underestimation of the potential of minority-led startups or misinterpretation of their pitches due to cultural differences in communication and presentation styles. These systemic issues perpetuate a cycle where minority founders find it increasingly difficult to break into the AI industry, leading to underrepresentation, stifled innovation, and the continuation of a technology landscape that may fail to address diverse communities' unique needs and perspectives.

Quiz

1. What is a primary benefit of engaging diverse stakeholders in AI governance?

    A) Reducing project costs

    B) Ensuring ethical AI development

    C) Limiting public involvement

    D) Speeding up AI deployment

2. Which method is NOT recommended for engaging stakeholders in AI governance?

   A) Public forums

   B) Ignoring stakeholder feedback

   C) Workshops and focus groups

   D) Online surveys

3. Why is building public trust important in AI governance?

   A) To increase AI adoption

   B) To avoid regulatory compliance

   C) To reduce the need for transparency

   D) To simplify technical jargon

4. Which stakeholder is typically involved in AI ethics discussions?

   A) Marketing team

   B) Data scientists

   C) Financial analysts

   D) Sales department

5. True or False: Stakeholder engagement should be a one-time task at the start of an AI project.

6. True or False: Engaging a broad range of stakeholders can help identify potential negative impacts of AI early on.

CHAPTER 13   THE POWER OF STAKEHOLDER ENGAGEMENT IN AI GOVERNANCE

7. True or False: Building public trust through transparency and inclusivity is not essential for AI governance.

8. Stakeholder engagement in AI governance fosters _____ and promotes the responsible use of AI.

9. Mapping stakeholders helps prioritize those with elevated responsibilities for increased and in-depth _____.

10. Diverse engagement methods include public forums, workshops, and _____ partnerships.

293

CHAPTER 14

Considering the Environmental Impacts of AI Systems

Considering the environmental impact of AI systems and ensuring they are developed and operated sustainably is incredibly careful planning and consideration.

AI systems have a power-hungry nature. Training large AI models, like those for natural language processing, requires massive energy. This energy often comes from sources that contribute significantly to carbon emissions. Reducing the energy footprint of AI is crucial for sustainability, as is ensuring that renewable sources are used in power data centers. The development and deployment of AI systems often rely on specialized hardware, which has its environmental costs during manufacturing. AI solutions requiring frequent hardware updates can exacerbate electronic waste (e-waste) problems. Designing for longevity, using responsibly sourced materials, and incorporating repair-friendly designs are necessary for reducing AI's contribution to e-waste. While training large models is energy-intensive, techniques like transfer learning can reduce the need to train from scratch every time. Optimizing algorithms for efficiency, choosing less computationally demanding models where fitting and compressing models for deployment can all lessen AI's environmental impact.

CHAPTER 14   CONSIDERING THE ENVIRONMENTAL IMPACTS OF AI SYSTEMS

Many AI applications rely on cloud services. Cloud providers are increasingly aware of their environmental impact and are investing in renewable energy sources. Choosing providers committed to sustainability can indirectly reduce the carbon footprint of your AI solutions. *Understanding the full lifecycle impacts of AI requires a sustainable approach that* extends beyond energy use during development. Consider the potential effects of AI-driven automation on resource consumption in other industries. For instance, while AI might optimize energy use in a factory, does it inadvertently drive increased production that negates efficiency gains? AI itself can be a powerful tool for sustainability. AI-powered systems can enhance energy grid efficiency, optimize resource management, model climate change scenarios, and aid in the discovery of eco-friendly materials. Using AI responsibly to address ecological challenges is crucial to its sustainability.

Transparency and Measurement → To track progress, companies need to establish clear metrics around the environmental impact of their AI systems. These include energy consumption, carbon emissions, and contribution to e-waste. Publicly reporting these metrics fosters accountability and encourages competition to improve sustainability.

Responsible Disposal and Recycling → When AI hardware reaches the end of its useful life, ensure responsible disposal and recycling practices to minimize e-waste pollution. Working with certified e-waste recyclers and exploring ways to repurpose outdated hardware reduces environmental harm.

Education and Awareness → Fostering a culture of sustainability within AI development teams is important. Developers need to understand the environmental consequences of their choices, from model architecture to hardware selection. Encouraging ongoing education about sustainable best practices is key for long-term progress. The sustainability of AI is not just a tech problem. Collaboration between AI developers, hardware manufacturers, cloud providers, and environmental scientists is needed. Working together can create holistic solutions that reduce environmental impact without compromising AI's potential benefits.

CHAPTER 14   CONSIDERING THE ENVIRONMENTAL IMPACTS OF AI SYSTEMS

Fostering a Culture of Continuous Improvement

As AI technology rapidly evolves, it is necessary to stay vigilant and reassess its environmental impact regularly. Adapting strategies, embracing ongoing innovation, and prioritizing eco-conscious solutions create a more responsible and sustainable future for AI development. While AI offers remarkable advancements, it comes with a hidden environmental cost. Training and running complex AI models, especially deep learning ones, requires massive computational power, consuming vast energy. This puts a strain on power grids and contributes to carbon emissions.

AI constantly evolves, with new algorithms, hardware, and applications emerging rapidly. As a result, the environmental impact of AI is not a static problem but a dynamic one. To ensure the long-term sustainability of AI development, we must adopt a culture of continuous improvement and vigilance. This involves regularly revisiting risk assessments to adapt to new threats, re-evaluating mitigation strategies based on the latest knowledge, and actively seeking innovative solutions. Prioritizing eco-conscious research and development will ensure that the AI revolution progresses in harmony with environmental responsibility rather than leaving a heavy ecological footprint for future generations to address.

When analyzing the impacts of AI systems, prioritizing energy efficiency in hardware must be a key factor. Designing energy-efficient AI-specific hardware (like Google's TPUs) is crucial. Optimizing hardware infrastructure within data centers, including power management and cooling systems, further reduces energy consumption. The foundation of energy-efficient AI lies in specialized hardware designed specifically for the unique demands of AI workloads. Traditional CPUs, while versatile, are not optimized for the massive parallel computations required by deep learning models. This is where AI accelerators like Google's Tensor

Processing Units (TPUs), NVIDIA's GPUs, and other emerging chip designs come in. These processors are engineered to excel at matrix multiplications and other operations fundamental to AI, delivering significantly higher performance per watt compared to general-purpose hardware. Beyond individual chips, optimizing the entire data center infrastructure plays a crucial role. This includes efficient power distribution, advanced cooling systems (potentially liquid or immersion cooling), and intelligent workload scheduling to maximize resource utilization.

The Power of Algorithmic Efficiency

While hardware optimization is crucial, reducing AI's energy footprint extends to the very heart of its software – the algorithms themselves. Even running on efficient hardware, overly complex AI models can still demand excessive computational resources. This is where algorithmic efficiency comes to the forefront. Deep neural networks can contain millions, even billions, of parameters (connections), and research has shown that many of these connections are redundant. As such, neural network pruning techniques strategically remove less important connections from a network. This results in a slimmer, more efficient model that achieves similar accuracy while requiring significantly less computation. The adage "more data is always better" does not always hold in the context of environmentally responsible AI. Indiscriminate data collection and storage inflates energy demands and contributes to digital waste. Collecting and storing excessive data inflates AI's environmental footprint unnecessarily. Strategies like selective data collection, efficient compression techniques, and responsible deletion of outdated data minimize this impact. Choosing data centers powered by renewable energy sources adds another layer of sustainability. Strategies for mindful data management include

- Performing selective data collection by critically evaluating what is essential for training and operating your AI model. Avoid collecting data habitually "just in case."

- Employ efficient data compression techniques to reduce data storage footprint without compromising its utility for AI purposes.

- Establish clear policies for responsible data retention and the secure disposal of data that is outdated, redundant, or no longer serves a valid purpose.

- **GO GREEN!** Partner with data center providers prioritizing renewable energy sources like solar or wind power. Choose locations with climates that allow for natural cooling mechanisms, further reducing the energy overhead of data storage.

Knowledge Distillation and Quantization

The knowledge distillation technique involves training a large, complex AI model (the "teacher") and then distilling its learned knowledge into a smaller, more efficient model (the "student"). The student model learns to mimic the outputs of the teacher, becoming a computationally lighter alternative suitable for energy-constrained environments. Neural networks typically use 32-bit floating-point numbers for calculations. Quantization reduces precision (to 16-bit or even 8-bit representations), lowering memory requirements and computational costs. While careful calibration is needed to maintain accuracy, quantization can offer significant optimization benefits, especially when used with knowledge distillation techniques.

CHAPTER 14   CONSIDERING THE ENVIRONMENTAL IMPACTS OF AI SYSTEMS

Knowledge distillation and quantization are complementary techniques that can be used together to significantly improve the efficiency and manageability of AI systems. Here is why this combination is powerful.

- **Reduced Model Size and Complexity** → Knowledge distillation creates a smaller student model that approximates the performance of the larger teacher model. Quantization further compresses the student model by reducing numerical precision. This combination results in a dramatically smaller model that is less computationally demanding.

- **Improved Inference Speed** → Smaller, quantized models execute much faster than their larger, high-precision counterparts. This is especially beneficial for real-time applications or those running on resource-constrained devices.

- **Lower Energy Consumption** → The reduced computational need for distilled and quantized models directly translates into lower energy consumption during training and inference. This is key for reducing the environmental impact of AI systems. The smaller model footprint enables the deployment of AI into edge devices, such as smartphones, wearables, and IoT sensors, with limited memory and processing power.

Example Scenario

- Start with a large, accurate, but computationally expensive AI model for image classification.

- Apply knowledge distillation to transfer knowledge from this model to a smaller, more efficient student model.

CHAPTER 14   CONSIDERING THE ENVIRONMENTAL IMPACTS OF AI SYSTEMS

- The smaller student model is then quantized to reduce its size and computational requirements.

- The result is a model that can be deployed on a mobile device while providing reasonable accuracy.

Understanding the Full AI System Lifecycle

A holistic approach requires a lifecycle assessment of AI systems. This means evaluating the environmental impact at every stage – from hardware manufacturing, energy-intensive model training, and ongoing deployment to eventual end-of-life disposal. To truly grasp an AI system's ecological implications, moving beyond a narrow focus on its training or operational phases is essential. A comprehensive lifecycle assessment (LCA) approach reveals the cumulative impact throughout each stage of an AI system.

Training large deep learning models, particularly in areas like natural language processing or computer vision, can take days, weeks, or even longer. The energy consumed during this phase is often the most significant contributor to an AI system's carbon footprint. Even after training, AI systems deployed in real-world applications continuously consume energy. The scale of deployment (e.g., embedded on millions of devices vs. running on a few servers) significantly impacts the overall energy demand.

Finally, responsible disposal or recycling of AI hardware is crucial to minimize e-waste and prevent harmful substances from ending up in landfills. Hardware design with easy disassembly and recoverability of valuable materials should be considered from the outset. As such, researchers continuously explore novel neural network architectures designed with efficiency in mind. Approaches like mobile-friendly networks (MobileNet, EfficientNet) and techniques like weight sharing or sparse connectivity patterns offer new avenues for reducing energy consumption without sacrificing performance.

CHAPTER 14   CONSIDERING THE ENVIRONMENTAL IMPACTS OF AI SYSTEMS

The Importance of Transparency

Tracking and openly disclosing metrics like energy consumption, carbon emissions, and water usage for AI models fosters accountability. Sharing successes and challenges in improving energy efficiency encourages others in the industry to do the same.

The first step toward transparency is defining clear and standardized metrics. This includes metrics for energy consumption (e.g., kilowatt-hours used during training), carbon emissions (both direct and indirect), water usage (especially for data center cooling), and potentially even e-waste generated through hardware turnover. Establishing these metrics allows for objective comparison and tracking progress over time. Organizations developing and deploying AI systems should regularly disclose their environmental impact metrics. This can be done through sustainability reports, dedicated websites, or partnering with third-party organizations that track ecological performance. Public reporting fosters accountability and allows stakeholders to make informed decisions about their interactions with AI-powered products and services.

Transparency is not just about individual responsibility; it's about driving the entire industry forward. By sharing knowledge on energy-efficient practices, challenges encountered, and innovative solutions, organizations can collaboratively set benchmarks and identify areas for improvement. This collective effort accelerates the development of sustainable AI practices and sets a positive example for other industries.

Partnering with policymakers, researchers, and industry leaders is essential for establishing industry-wide standards that measure AI's environmental footprint and drive the development of best practices for sustainability.

CHAPTER 14   CONSIDERING THE ENVIRONMENTAL IMPACTS OF AI SYSTEMS

Integrating Sustainability into AI Design

Environmental considerations should be built into AI development from the outset. Prioritizing energy efficiency and sustainability as core design objectives guides decisions throughout creation. Moving beyond reactive measures, making environmental considerations a core pillar of AI development from its earliest stages is paramount. This proactive mindset shift can drive eco-conscious innovation and limit the need for after-the-fact mitigation efforts. From the initial conceptualization, prioritize energy efficiency as a non-negotiable design goal to ensure you are designing AI for efficiency and sustainability. Explore algorithms and architectures with low computational complexity. Evaluate potential trade-offs between performance and energy consumption holistically, considering future implications.

Whenever possible, favor smaller, more efficient AI models that achieve the desired level of functionality without excessive resource hunger. This is particularly relevant for edge devices, wearables, and mobile applications where hardware limitations and battery life are major factors. Explore techniques to reduce the energy consumption of the model training process itself. Strategies include utilizing renewable energy sources for power-hungry computations, scheduling training during off-peak hours when the grid is less strained and experimenting with distributed training models. Establish metrics to track and communicate energy efficiency gains throughout the development process. Measure and report on indicators like carbon footprint, power usage effectiveness (PUE), and water usage of AI systems. Share these measures openly to promote industry-wide accountability.

Expand traditional AI development workflows to include environmental impact as a key evaluation metric. Challenge the notion that accuracy or speed is the sole determinant of success. Reward innovation that reduces environmental impact alongside traditional performance gains.

CHAPTER 14   CONSIDERING THE ENVIRONMENTAL IMPACTS OF AI SYSTEMS

Summary and Thoughts

The rapid advancement of AI technology offers tremendous opportunities for innovation and societal progress. However, it is essential to acknowledge the hidden environmental costs associated with the energy-intensive nature of AI development and deployment. Focusing on energy efficiency and sustainability throughout every stage of an AI system's lifecycle is crucial for responsibly harnessing AI's full potential.

AI-based models and solutions lie in a multifaceted approach. This includes optimizing AI-specific hardware, developing computationally efficient algorithms, prioritizing mindful data management, and embracing green data center practices. Transparency in tracking and reporting environmental metrics is critical for accountability, and proactive research focused on sustainable AI innovation will pave the way for the future.

Ultimately, integrating environmental considerations into the core of AI design will foster a culture of continuous improvement. It is no longer enough to mitigate existing AI impact after the fact simply; the goal must be the development of inherently sustainable AI systems. This collaborative effort, where industry leaders, policymakers, and researchers work together, is key to ensuring that the AI revolution progresses harmoniously with our planet's well-being.

Quiz

1. What is a major environmental concern associated with AI systems?

    A) High software costs

    B) Energy consumption

    C) Data redundancy

    D) Limited computational power

CHAPTER 14   CONSIDERING THE ENVIRONMENTAL IMPACTS OF AI SYSTEMS

2. Which practice helps reduce the environmental impact of AI systems?

    A) Ignoring hardware updates

    B) Using renewable energy sources

    C) Reducing transparency

    D) Increasing electronic waste

3. What is the significance of algorithmic efficiency in AI?

    A) It increases energy consumption

    B) It reduces computational costs

    C) It minimizes data storage

    D) It maximizes e-waste

4. Which of the following does NOT contribute to AI's environmental impact?

    A) Specialized hardware

    B) Frequent software updates

    C) Use of non-renewable energy

    D) High data quality

5. True or False: The development of AI systems has no significant environmental impact.

6. True or False: Integrating sustainability into AI design can help mitigate its environmental impact.

7. True or False: Transparency in AI operations does not affect its environmental footprint.

CHAPTER 14   CONSIDERING THE ENVIRONMENTAL IMPACTS OF AI SYSTEMS

8. To reduce environmental impact, AI systems should use energy from _____ sources.

9. AI systems can exacerbate electronic waste problems if not designed for _____.

10. Algorithmic _____ is crucial for minimizing the energy footprint of AI systems.

CHAPTER 15

Developing the Protocols for Rapid Response in Case of an AI-Related Crisis

Artificial intelligence (AI) has become an integral part of our lives, with applications ranging from healthcare and finance to transportation and customer service. However, this increased reliance brings a heightened risk of AI-related crises. These could manifest as malfunctions leading to critical system failures, biased or discriminatory outputs affecting large populations, or even the emergence of AI systems that act against their intended purpose. There is an urgent need for proactive AI crisis management. Traditionally, crisis management has often been reactive as organizations respond to incidents as they occur. However, this approach is inadequate when dealing with AI crises' complexity and potential scale. A malfunctioning AI system has the potential to cause widespread damage in a matter of seconds or minutes, far faster than traditional crises might unfold. Proactive AI crisis management means anticipating potential failure points, developing mitigation strategies in advance, and having protocols for rapid and decisive action. This preparedness minimizes damage, protects public trust in AI, and safeguards the organization's reputation. By shifting the mindset from "if" a crisis occurs to "when" it does, organizations are better equipped to deal with even the most unforeseen AI-related events. Moving from a reactive to a proactive approach to AI crisis management is essential. Preemptive measures are far more effective than scrambling to contain damage after an incident occurs. If your organization utilizes AI, you must have a comprehensive crisis management plan specifically designed for AI-related scenarios. The following crisis management strategy is recommended as a starting point for your plans.

Anticipatory Risk Assessment and Management

The first step is a thorough risk assessment. *AI Risk Management is covered in Chapters 3 and 5 of this book.* Organizations need to identify potential failure points within their AI systems, evaluate the severity of the possible consequences, and prioritize mitigation strategies. This assessment should not be a one-time but an ongoing process, constantly adapted as AI systems evolve and potential risks shift.

Developing Rapid Response Protocols

Rapid response is paramount in an AI crisis to minimize damage and public distrust. Protocols should include a clear chain of command, identifying decision-makers and their roles. Predefined communication channels, both internal and external, should be in place to streamline information flow and avoid confusion. Every second of delay can worsen the situation and escalate the fallout. Rapid response protocols are the backbone of effective AI crisis management. These protocols should include

CHAPTER 15   DEVELOPING THE PROTOCOLS FOR RAPID RESPONSE IN CASE OF AN AI-RELATED CRISIS

Clear Chain of Command. Predetermined decision-making hierarchy outlining who has the authority to initiate shutdown procedures, make critical judgment calls, and communicate with stakeholders. This avoids confusion and bottlenecks during high-stress moments.

Communication Channel. Designated communication channels for both internal and external use. Internal channels should streamline coordination between technical teams, management, and legal advisors. External channels must be established to communicate with affected parties, the public, the media, and relevant regulatory bodies.

Swift Data Preservation. Protocols should address quickly securing logs, system snapshots, and any data relevant to forensic analysis. This is essential for understanding the root cause of the crisis and improving systems down the line.

Triage and Prioritization. In large-scale incidents, there might be cascading effects. Protocols should guide teams to prioritize the most critical systems or functions needing immediate shutdown or isolation, ensuring that safety and core functionality take precedence.

Simulation and Practice. Even the best-laid plans are worthless if not practiced. Regular simulations of AI crisis scenarios involving all relevant personnel help test the readiness of protocols, fine-tune communication, and reveal potential gaps in the overall response strategy.

Emergency Shutdown Procedures

Safe and controlled shutdown procedures are critical. These need to be meticulously designed and evaluated well in advance. A simple "power-off" solution might not be feasible in complex AI systems. The protocols should address isolating malfunctioning components, severing external output channels, and preserving data for forensic analysis. Safe, controlled shutdown procedures are the cornerstone of minimizing damage during an AI crisis. Designing them requires careful consideration and rigorous testing.

Controlled vs. Abrupt Shutdown. While a simple power-off might be appropriate for some systems, complex AI implementations often require more nuanced approaches. Severing data flows, isolating specific modules, or enacting a staged shutdown sequence might be necessary to prevent further harm.

Data Integrity. Shutdown procedures must balance stopping a malfunctioning system and preserving vital data. Provisions should be made for capturing critical system state information, logs, and even memory snapshots that will be crucial for later forensic analysis.

Redundancies and Safety Checks. In safety-critical applications, shutdown procedures might include built-in redundancies and multiple confirmation steps to reduce the risk of human error during a tense situation.

External Interfaces. Mapping all output channels of an AI system is essential. Shutdown procedures need to account for how to sever communication with external hardware (like machinery), Internet connections, or even other systems to prevent the crisis from cascading.

Reversibility. Depending on the nature of the crisis, it is worth considering procedures to "restart" or roll back the AI system to a safe state after dealing with the immediate situation. This enables faster recovery and aids with the investigation.

Communication Transparency

In a crisis, transparent and timely communication with stakeholders – the public, regulators, and affected individuals – is vital. Proactive communication builds trust and minimizes the potential for panic or misinformation. Organizations need designated spokespeople prepared to explain the situation in clear, understandable language and address evolving concerns. In the aftermath of an AI crisis, public trust is paramount. Transparent and timely communication with stakeholders – including affected individuals, regulators, the media, and the public – can significantly influence how the crisis is perceived and managed. Here's why transparency is essential:

Managing Public Perception. Proactive, clear communication fosters a sense of control and minimizes fears that might escalate due to misinformation or the lack of updates. Organizations need designated spokespeople to explain complex technical issues in accessible language to a broad audience.

Containing the Crisis. Transparency builds trust, which can help contain the crisis. Being up-front about the scope of the problem, initial findings, and steps taken for resolution can prevent rumors and speculation.

Learning from Incident Transparency opens opportunities for constructive feedback from experts and the public. This feedback can be invaluable in the post-crisis review, leading to enhanced protocols and system safeguards.

Ethical Responsibility. Organizations that depend on AI have an ethical responsibility to be transparent, especially when malfunctions cause harm. This includes acknowledging mistakes, offering apologies when called for, and outlining steps taken to prevent similar incidents in the future.

Collaborative Incident Response

Collaboration among organizations, industry experts, and even government agencies may be needed depending on the scale of an AI crisis. Pre-established networks and partnerships facilitate the rapid sharing of information, resources, and expertise in a time when such coordination can prove lifesaving. The scope and complexity of certain AI crises might exceed the capabilities of a single organization to manage effectively. In such scenarios, swift and coordinated collaboration becomes vital. Here is why the swift and collaborative response is imperative and how to ensure adequate incident-managed response:

Resource Sharing. Organizations might need to pool technical expertise, computing resources, or specialized tools during a large-scale crisis. Pre-established collaboration networks facilitate the rapid mobilization of these resources when time is of the essence.

Industry-Wide Coordination. Collaboration across an industry sector may be needed when an AI crisis reveals systemic vulnerabilities. Companies can share anonymized data and analysis to identify common failure patterns, accelerating the development of solutions and preventive measures.

Government Partnerships. Certain crises might necessitate public-private collaboration. Government agencies can offer regulatory insights, additional technical resources, legal authority, or assistance in communicating to the public with a unified voice.

Pre-established Networks. Building these collaborative networks cannot happen in the heat of a crisis. Organizations should proactively engage with peers, industry bodies, relevant academic institutions, and even potential government partners to formalize communication channels and establish protocols for sharing information in times of emergency.

Forensic Analysis and Post-crisis Review

Each AI crisis becomes a learning opportunity. Detailed forensic analysis is essential to understand the root cause of the malfunction. Findings should then feed into refining protocols, enhancing system design, and improving risk assessment models. Sharing lessons learned across the broader AI community fosters industry-wide resilience. *The digital crime scene should be treated much like a traditional crime scene, and the AI system and its surrounding environment should* become the focal point of a forensic examination. Investigators meticulously gather data, including system logs, code snapshots, input and output records, network traffic, and user interaction histories. This data offers clues about what occurred in the moments leading up to, during, and after the AI crisis. Forensic analysis involves more than simply identifying errors. It is about tracing the chain of events back to their origin. This could include dissecting flawed algorithms, uncovering biases in training data, pinpointing misconfigurations, or even identifying points where external interference might have occurred. Thorough analysis should not be confined to the technical realm but expanded across your organization. Human factors, i.e., inadequate training, confusing interfaces, or inappropriate reliance on the AI system's output, might have played a significant role.

Investigators and forensics must evaluate the entire socio-technical system within which the AI operates. As such, consider systemic vulnerabilities. An in-depth analysis can reveal vulnerabilities that extend beyond the specific malfunctioning system. A crisis often exposes flaws in organizational processes, risk identification, control procedures, oversight mechanisms, or the broader regulatory landscape. These findings are vital for bolstering resilience across entire industries against future incidents. True prevention depends on sharing anonymized lessons learned from forensic analysis. Industry associations, academic institutions, and governmental bodies can function as clearinghouses for such knowledge. By understanding common failure patterns and systemic weaknesses, the entire AI community can implement proactive safeguards and raise the standard of safety and reliability.

Ethical Considerations in AI Crisis Management

AI crisis management needs to be deeply intertwined with ethical considerations. Decisions made during the crisis, especially trade-offs between safety and efficiency, should align with an organization's ethical AI framework. This involves preemptively defining priorities and guiding principles for crisis scenarios. Organizations utilizing AI should

CHAPTER 15 DEVELOPING THE PROTOCOLS FOR RAPID RESPONSE IN CASE OF AN AI-RELATED CRISIS

have a predefined ethical framework guiding their actions under normal conditions, especially during a crisis. This framework should clearly articulate priorities such as:

Prioritizing Human Well-being. The safety and well-being of individuals potentially affected by the AI crisis should always precede other concerns. This may require difficult trade-offs regarding system shutdown procedures or information disclosures.

Non-Discrimination. Crisis response protocols cannot introduce or exacerbate biases and discrimination against any group of individuals. This implies scrutiny of decisions that impact data collection, resource allocation, or communication strategies during the crisis.

Transparency and Accountability. Even during a rapidly evolving situation, an organization has an ethical duty to explain its actions and decisions to the affected parties and the public. This fosters trust and demonstrates a commitment to accountability.

Respecting Privacy. Depending on context, privacy concerns may need to be balanced against safety during a crisis. However, organizations should strive to minimize any infringement on personal data, adhering to privacy laws and ethical norms.

Anticipating Ethical Dilemmas. It is vital to preemptively consider the ethical dilemmas that may arise during AI crises specific to the industry or application. This could involve

Resource Allocation. If AI-driven systems are used in a healthcare setting to triage patients, what are the ethical considerations when demand exceeds supply? Will the algorithms inadvertently prioritize or disadvantage any group?

Triage of Risk. When faced with an escalating AI malfunction, will prioritization be based purely on technical feasibility, or will potential human harm also be a critical factor in decision-making?

CHAPTER 15 DEVELOPING THE PROTOCOLS FOR RAPID RESPONSE IN CASE OF AN AI-RELATED CRISIS

Liability. Determining responsibility in an AI crisis can be exceedingly complex. Pre-established principles are needed to address questions of liability concerning system designers, developers, deployers, and even users who may have interacted with the system.

Incorporating Diverse Perspectives → Ethical considerations during crises should not be an afterthought or limited to technical experts. Organizations would benefit from

- **Multi-stakeholder Input** → Forming advisory groups that include ethicists, social scientists, legal experts, and representatives from potentially affected communities. This ensures a broader spectrum of concerns is considered.

- **Independent Oversight** → External oversight bodies can provide much-needed objectivity in reviewing crisis protocols and their ethical implications. This can help mitigate blind spots that might exist within the organization itself.

Continuous Evaluation and Adaptation → As AI technology evolves, ethical considerations surrounding its crisis management must evolve. This means

- **Re-evaluation of Frameworks** → Ethical principles and priorities must be revisited regularly to remain relevant in emerging social and technological complexities.

- **Flexibility in Approach** → Ethical decision-making should not be rigid. Protocols should allow for adaptation and re-evaluation in the specific context of an evolving crisis.

CHAPTER 15   DEVELOPING THE PROTOCOLS FOR RAPID RESPONSE IN CASE OF AN AI-RELATED CRISIS

Continual Evolution and Vigilance

The AI landscape is constantly evolving. The key takeaway is that AI crisis management needs to be similarly dynamic. Protocols cannot be static documents. They require regular reviews, updates, and simulations to ensure their effectiveness in the face of emerging technologies and the increasingly complex threats they may pose.

Monitoring the Tech Landscape

The field of AI is advancing at a breathtaking pace. New architectures, increased reliance on large language models, and novel AI applications constantly emerge, posing new and unforeseen risks. Organizations need dedicated teams tasked with horizon scanning, primarily identifying potential threats and future-proofing crisis management plans.

Proactive Risk Reassessment

Risk assessments should not be a one-and-done exercise. Regularly re-evaluating threats and vulnerabilities is essential as AI systems evolve, becoming more interconnected and interacting with complex real-world environments. Crisis management plans should be adjusted accordingly, with an eye toward new risks that may have arisen.

Stress Testing and Simulations

Testing crisis readiness should not be limited to technical aspects. Your stress testing and simulations must also include ethical stress testing. Simulations can put decision-makers in hypothetical scenarios with complex ethical trade-offs, forcing them to confront the potential implications of their actions in a safe environment. Finally, and most importantly, adversarial stress testing and simulation are vital. Incorporating "red team" exercises can help expose critical vulnerabilities that might have been missed. This helps anticipate malicious actors exploiting AI weaknesses during a crisis.

Engaging with Researchers

CHAPTER 15 DEVELOPING THE PROTOCOLS FOR RAPID RESPONSE IN CASE OF AN AI-RELATED CRISIS

AI crisis prevention is not only an industry responsibility. Building strong ties with the academic research community can offer valuable insights and early warnings. Companies can fund research projects focusing on AI failure modes, safety analysis, and developing more robust AI architectures. Promoting standards development and actively participating in the development of industry-wide AI standards for safety, reliability, and explainability will benefit everyone. Companies should not see these standards as constraints but as opportunities for preemptive risk mitigation. AI crises do not respect borders nor act with prejudice to a specific country or region. Fostering international collaboration on AI safety protocols is vital. This includes

- **Sharing Incident Data** → Creating anonymized and secure data-sharing platforms where organizations worldwide can learn from each other's AI crises.

- **Joint Exercises** → Simulated multinational responses to AI incidents can help streamline communication, identify areas of friction, and build pre-existing collaboration mechanisms for when an actual global-scale crisis occurs.

The Importance of Explainability AI systems is often opaque "black boxes." Understanding the reasoning behind an AI system's output becomes paramount in a crisis. Promoting investment in explainable AI (XAI) techniques allows for quicker diagnosis and resolution of malfunctions. It is crucial to remember that AI exists within a wider socio-technical system, and vigilance includes

- **Upskilling and Re-tooling the Workforce** → Continuous training ensures employees who interact with AI systems understand their strengths and limitations, empowering them to recognize potential signs of malfunction early on.

CHAPTER 15   DEVELOPING THE PROTOCOLS FOR RAPID RESPONSE IN CASE OF AN AI-RELATED CRISIS

- **Public Education** → Fostering basic AI literacy among the public helps manage expectations, builds trust, and prepares society to participate actively in the conversation about ethical AI crisis management.

Summary and Thoughts

The increasing use of AI necessitates proactive crisis management. Organizations should continually conduct risk assessments, anticipating potential AI failures to tailor their response. Rapid response protocols with clear decision-making chains and communication strategies are crucial. Emergency shutdown procedures must be well-defined, allowing for the safe containment of AI malfunctions. During a crisis, transparent and timely communication with stakeholders builds trust and prevents misinformation. Collaboration between organizations and experts might be essential for an effective response, depending on the scale.

Every AI crisis should be thoroughly analyzed, with findings used to improve protocols, system design, and overall risk management across the industry. Ethical considerations must guide AI crisis management decisions. Organizations need pre-established ethical frameworks to navigate difficult trade-offs. The dynamic nature of AI demands that crisis management plans be constantly reviewed, updated, and tested to ensure their effectiveness as threats and technologies evolve.

CHAPTER 15   DEVELOPING THE PROTOCOLS FOR RAPID RESPONSE IN CASE OF AN AI-RELATED CRISIS

Quiz

1. What is a key component of rapid response protocols for AI-related crises?

    A) Ignoring potential risks

    B) Anticipatory risk assessment

    C) Delaying communication

    D) Avoiding forensic analysis

2. Which aspect is NOT part of AI crisis management?

    A) Emergency shutdown procedures

    B) Collaborative incident response

    C) Ignoring post-crisis reviews

    D) Ethical considerations

3. What is crucial for effective communication during an AI crisis?

    A) Minimizing stakeholder involvement

    B) Transparency and timeliness

    C) Hiding information

    D) Delaying responses

4. Which practice ensures continuous improvement in AI crisis management?

    A) Ignoring feedback

    B) Regular audits

    C) Reducing stakeholder communication

    D) Minimizing transparency

CHAPTER 15 DEVELOPING THE PROTOCOLS FOR RAPID RESPONSE IN CASE OF AN AI-RELATED CRISIS

5. True or False: Emergency shutdown procedures are not necessary in AI crisis protocols.

6. True or False: Collaborative incident response is crucial for managing AI-related crises effectively.

7. True or False: Post-crisis reviews help organizations learn and improve their crisis management protocols.

8. Rapid response protocols should include _____ risk assessment and management.

9. Effective communication during a crisis must be _____ and transparent.

10. Forensic analysis and _____ reviews are essential for improving AI crisis management.

CHAPTER 16

Capacity Building for AI Actors

A comprehensive approach to skills development for stakeholders and actors is essential to ensure AI technologies' effective and responsible deployment. This includes targeted training programs for regulators to understand AI's capabilities and potential risks, empowering them to create informed policy frameworks. Developers should receive continuous education on ethical AI design, bias mitigation, and explainability techniques. Moreover, initiatives tailored to educate end-users about AI's benefits, limitations, and safe usage practices will promote public trust and informed engagement with this transformative technology.

Developing Skills and Knowledge Among AI Actors and Stakeholders

Developing skills and knowledge among stakeholders, including regulators, developers, and end users, to effectively understand and manage AI technologies is often "easier said than done." The rapid pace of AI innovation has outpaced the understanding of this technology among key stakeholders and actors. Regulators grapple with complex technical concepts, developers can be unaware of social implications, and end users may lack awareness of both the potential and the risks that AI poses. This knowledge gap can hinder responsible AI implementation and trust in the technology. Skill development ensures effective, ethical, and safe AI use. Informed regulators can create governance frameworks that balance innovation and public safeguards. Knowledgeable developers who focus on governance create AI systems with fairness, explainability, and data and privacy protection built in from the start. Empowered end users can make conscious choices about AI interactions, avoiding unintended harm and demanding accountability.

Regulators require AI literacy to understand capabilities, biases, governance, and risks. Training programs should cover technical fundamentals, ethical implications, and case studies of successful and problematic AI deployments. This knowledge empowers them to create regulatory frameworks that protect the public without stifling innovation.

Well-Governed AI Design for Developers

Developers need ongoing education about ethical AI principles. These include techniques to identify and mitigate biases in datasets, build transparency and explainability into models, and prioritize privacy protection. Embedding ethical considerations throughout development leads to more reliable and trustworthy AI systems.

Explainable AI (XAI) is essential for both developers and regulators. Developers should master XAI techniques to create models that justify their reasoning, enabling debugging and ethical troubleshooting. Regulators need to understand XAI to assess the fairness and trustworthiness of the AI systems they oversee. Explainable AI (XAI) addresses complex models' "black box" nature. It provides techniques for developers to open their AI systems, allowing them to understand why models arrive at certain decisions. This understanding is vital for several reasons: debugging errors, ensuring fairness, building trust with users affected by AI decisions, and allowing regulators to verify that models comply with guidelines.

AI systems often inherit biases from the data they are trained on or reflect the unconscious biases of their developers. To counter this, developers must be actively trained to identify and mitigate such biases throughout the design process. Techniques like dataset balancing, fairness testing tools, and adversarial de-biasing allow developers to detect and minimize discriminatory outputs from their AI models. Additionally, a deep understanding of data governance and privacy protection is essential, ensuring user data is respected and used responsibly.

Fostering collaboration between developers and experts in fields like social sciences, ethics, and law brings multifaceted perspectives to AI design. This interdisciplinary exchange helps developers anticipate societal impacts and proactively create more inclusive and responsible solutions. Collaboration between AI developers and experts from various disciplines can significantly enhance the social responsibility of AI products. Social scientists can pinpoint potential areas where biases might creep in, ethicists can guide in balancing efficiency with fairness, and legal experts can advise on regulatory compliance. This multidisciplinary approach gives developers a broader perspective, reducing the risk of unintended negative consequences and ensuring their AI systems uphold societal values alongside technical excellence.

AI Literacy for End Users

Empowering end users starts with basic AI literacy. Educational initiatives should clarify how AI systems work and their benefits and risks, such as bias or privacy intrusion. Users should also be informed of their rights concerning AI-driven decisions affecting their lives. AI literacy for typical users is critical for fostering informed interactions with this increasingly influential technology. Educational programs must go beyond basic awareness, equipping users with the knowledge they need to participate actively in the AI-driven world. Demystifying AI models and systems helps users understand its strengths and limitations. Organizations that build AI systems and programs should offer simplified explanations of fundamental AI concepts for how models are trained, different types of AI techniques (e.g., machine learning, natural language processing), and common applications that users may encounter daily. Highlighting positive use cases of AI can reduce anxiety and promote adoption. Users should know how AI can enhance healthcare, financial literacy, education, customer service, and environmental sustainability. Illustrating tangible benefits helps balance awareness of risks with recognition of AI's value. Alongside benefits, users need to be aware of AI's potential pitfalls. Educational resources should address issues like the perpetuation of biases in datasets, the risks of deepfakes and misinformation, and the importance of safeguarding personal data. This understanding enables users to exercise caution and advocate for their rights.

Users should not feel helpless when facing AI systems and programs. Knowledge about their rights regarding AI-driven decisions is empowering. Education should inform users of existing or emerging regulations related to data privacy protection, the right to explanations about decisions impacting them, and avenues for voicing concerns about the potential misuse of AI. Non-technical users have a wealth of resources to start their AI education journey. Free online courses (like Elements of AI or AI for Everyone) offer introductory overviews. Websites, blogs, and interactive AI experiments provide diverse ways to learn about concepts and applications. For those who prefer reading, non-technical books explore AI's impact and ethical considerations. No matter your learning style, there is an accessible resource to help you understand the world of AI.

Author's hint This book is an excellent gift for users seeking to expand their understanding of AI Systems and Programs.

CHAPTER 16   CAPACITY BUILDING FOR AI ACTORS

AI Literacy for Business Executives

AI literacy for executives entails developing a working knowledge of AI technologies, recognizing their potential impact across various business functions, and understanding the ethical considerations and risks involved. Executives gain the necessary insights to lead their organizations into the future by actively pursuing AI literacy. They can identify opportunities for AI adoption, align AI initiatives with business goals, and foster a culture of responsible AI development, ensuring their organizations thrive in the age of intelligent automation.

In today's technology-driven world, Artificial Intelligence (AI) is rapidly transitioning from a futuristic concept to a business imperative. AI can transform industries, optimize business operations, and enhance decision-making. However, harnessing this potential requires AI literacy among executives. It's no longer enough to delegate AI to the IT department; business leaders must understand its capabilities and limitations to make strategic, forward-thinking decisions. Artificial Intelligence (AI) is transforming industries at a breathtaking pace. From streamlining operations to revolutionizing customer experiences, AI is reshaping the competitive landscape. In this rapidly changing environment, AI literacy for business executives is no longer a luxury; it's necessary for survival and long-term success. Understanding the AI landscape sets the foundational principles for technology-driven organizations. AI literacy transcends mere technical knowledge. Business leaders need a solid understanding of various AI concepts, techniques (e.g., machine learning, natural language processing), and how AI can create value across the organization. This understanding empowers executives to identify high-potential AI use cases, evaluate solutions, and strategically align AI initiatives with overall business goals. AI-literate executives can make informed decisions about adopting, integrating, and scaling AI solutions. They can critically assess AI vendors and proposals, weigh the potential risks and benefits, and ensure

CHAPTER 16   CAPACITY BUILDING FOR AI ACTORS

ethical and responsible implementation aligned with the company's values. This informed approach steers businesses towards AI investments that deliver real, measurable returns.

A recent conversation with several senior executives spawned the following question. "How do we unlock the competitive advantages AI provides?"

AI can provide businesses with significant competitive advantages in many ways. Process automation can drive efficiency, reduce costs, and save valuable employee time for higher-value tasks. AI-powered data analysis can reveal insights, driving better customer targeting, product development, and market prediction. Personalized customer experiences powered by AI boost loyalty and customer satisfaction. I provided the following responses to their question:

- "**Process Automation** → Implement AI solutions to streamline repetitive, time-consuming tasks across various departments. This frees employees to focus on strategic initiatives and improves operational efficiency, leading to cost savings."

- "**Data-Driven Insights** → Leverage AI-powered analytics to extract actionable insights from vast customer data, market trends, and internal operations. These insights, for example, can enable more precise targeting, informed product development, and proactive risk mitigation."

- "**Personalized Customer Experiences** → Utilize AI for real-time customer interaction analysis, recommendation engines, and intelligent chatbots. This enhances customer satisfaction, boosts loyalty, and drives sales growth."

CHAPTER 16   CAPACITY BUILDING FOR AI ACTORS

There are more ways to unlock the competitive advantages of AI, and I could dedicate an entire chapter to AI use cases and the benefits of AI realization. However, this book is about AI governance and risk management.

Convincing My Board of Directors and Senior Leadership Team to Use AI

The business landscape is rapidly evolving and driven by disruptive technologies such as artificial intelligence. Competitors leverage AI to gain significant advantages, from streamlining operations to personalizing customer experiences. Maintaining a competitive edge and ensuring future success demands that your organization proactively embrace AI, transforming potential disruption into opportunity. Obtaining tangible value with focused investments is the primary goal. Adopting AI is not about chasing every technological trend. Your organization must prioritize AI initiatives with the highest potential return on investment and, through targeted use cases, your organization can accomplish the following:

- Automate key processes to reduce costs, free up valuable employee time, and leverage AI-powered data analysis to drive better decision-making and enhance customer engagement with personalized AI solutions.

- Investing in future readiness and innovation has benefits beyond immediate ones. AI adoption signals to the market and stakeholders that you are a forward-thinking organization. It motivates you to adapt quickly to future trends and capitalize on emerging opportunities.

- Investing in AI fosters a culture of innovation, attracts top talent, and solidifies your reputation as an industry leader.

AI adoption requires careful planning. You should propose a phased approach, starting with pilot projects in areas where AI can deliver measurable results. Prioritize the ethical use of AI, implement data governance and safeguards for data privacy, and invest in upskilling your workforce to ensure a smooth transition. This measured and strategic approach maximizes potential benefits while proactively mitigating risks.

Recent studies show the positive benefits of AI-literate Boards and leadership teams and how they create a culture conducive to innovation. They encourage experimentation, empower employees to explore new AI applications, and foster a data-driven mindset across the organization. This proactive approach positions businesses to quickly adapt to emerging opportunities and disruptions, keeping them ahead of the competition in a constantly evolving market.

In contrast, premature AI adoption without adequate preparation can lead to wasted resources, project failures, and even reputational damage. Rushing into AI without clearly defined goals, lacking AI-ready data infrastructure, insufficient technical expertise, or ethical blind spots can result in poorly scoped projects that deliver minimal value. Unaddressed biases within training data or flawed algorithms can perpetuate discrimination and harm customer trust. Investing in AI readiness – including upskilling personnel, establishing data governance processes, and developing a robust ethical framework – is essential before substantial AI projects are undertaken. This measured approach ensures AI investments are strategic and sustainable, providing long-term value rather than costly setbacks.

Summary and Thoughts

Developing skills and knowledge in the age of AI and as AI permeates various industries, all stakeholders must build relevant knowledge and skillsets. This includes technical developers who need expertise in AI algorithms, data science, and programming. Business leaders require the ability to identify AI opportunities and integrate these technologies into strategic planning. End-users and the wider workforce also need basic AI literacy to understand AI system outputs, collaborate effectively with AI, and adapt to constantly changing job requirements driven by technological advancements. From the classroom to professional development, education significantly prepares individuals for success in this new technological landscape.

Why should companies prioritize AI investment?

Investing in AI systems offers businesses a significant competitive edge. AI can automate routine tasks, freeing employees from higher-value work and increasing efficiency and productivity. In addition, AI-driven insights can enhance decision-making, optimize processes, and unlock new revenue streams. AI facilitates the development of personalized products and services, fostering deeper customer relationships and brand loyalty. Importantly, companies that lag in AI adoption risk becoming obsolete and outpaced by competitors embracing this transformative technology.

AI is no longer an optional technology but a driving force shaping industries. Companies that fail to adapt to this AI-driven landscape risk being left behind. Being AI-ready means proactively identifying potential AI applications within your business. It involves investing in AI infrastructure, developing AI-related skills among your workforce, and fostering a culture that embraces data-driven decisions and innovation. Being AI-ready cultivates a mindset of agility within organizations. They are better equipped to respond to disruptions, pivot strategies, and capitalize on emerging AI-powered trends. AI readiness promotes a proactive approach to innovation, ensuring long-term success in a dynamic environment.

CHAPTER 16   CAPACITY BUILDING FOR AI ACTORS

Quiz

1. What is essential for capacity building among AI actors?

    A) Ignoring skill development

    B) Continuous training and education

    C) Reducing stakeholder involvement

    D) Avoiding new technologies

2. Which group should receive specialized training in ethical AI development?

    A) Sales teams

    B) Data scientists

    C) Marketing teams

    D) Financial analysts

3. Why is AI literacy important for business executives?

    A) To increase data redundancy

    B) To make informed decisions

    C) To reduce innovation

    D) To limit AI adoption

CHAPTER 16   CAPACITY BUILDING FOR AI ACTORS

4. What is a key focus area for developing AI literacy among end users?

   A) Ignoring privacy concerns

   B) Promoting AI transparency

   C) Reducing stakeholder feedback

   D) Avoiding ethical considerations

5. True or False: Capacity building for AI actors only involves technical training.

6. True or False: AI literacy for business executives helps them understand the strategic value of AI.

7. True or False: Continuous education and training are not necessary for maintaining AI skills and knowledge.

8. Capacity building for AI actors involves developing _____ and knowledge among stakeholders.

9. AI literacy for end users includes understanding the _____ of AI technologies.

10. Business executives need AI literacy to make _____ decisions regarding AI adoption.

CHAPTER 17

Intellectual Property Rights with AI Technologies

The legal landscape surrounding intellectual property (IP) and AI-generated content is still evolving. Currently, copyright protection for AI-created works is unclear in many jurisdictions. While human creators involved in developing and training AI systems may have claims to ownership, AI itself is not recognized as an "author" in most copyright laws. This raises questions about ownership and infringement when AI content is derivative of copyrighted materials used in training data. On the other hand, some argue that significant human input in crafting prompts and curating outputs justifies copyright protection for AI-generated creative content. Businesses and organizations using AI should know these uncertainties and consider potential legal implications when deploying AI systems and utilizing their outputs.

CHAPTER 17   INTELLECTUAL PROPERTY RIGHTS WITH AI TECHNOLOGIES

Addressing the Challenges Related to the Ownership and Sharing of AI-generated Content and the Technologies Themselves

A core issue is determining who, if anyone, can claim ownership of AI-generated content. Traditional copyright laws often require a human author. Can the AI developer, the person who inputs prompts, the owner of the training data, or the AI model or algorithm itself be considered the rightful owner? This legal ambiguity creates uncertainty about licensing and the fair use of AI-generated materials. Large AI models are often trained on massive datasets that might include copyrighted works like text, images, or code. Who owns the rights to content created by AI that has learned from potentially copyrighted sources? Questions about the legitimacy of using this data and the ownership implications for derived outputs remain a complex legal battleground. As you can see – there are many questions but very few answers to the "Authorship Conundrum."

Businesses investing in AI development want to protect their intellectual property. However, the collaborative and often open-source nature of AI innovation makes delineating clear ownership of algorithms, models, and the data they are built on difficult. Striking a balance between protecting investment and fostering innovation is essential. Developing standardized labeling systems or metadata identifying AI contributions can help establish accountability and trust.

Human Authorship vs. Machine Agency

The heart of the ownership debate lies in whether a machine can be considered an author. Traditional copyright laws were designed to protect the intellectual creations of humans. AI systems that generate seemingly

CHAPTER 17   INTELLECTUAL PROPERTY RIGHTS WITH AI TECHNOLOGIES

original content challenge these long-held assumptions. Can AI express the necessary creativity and independent thought to be considered its author, or is it simply a sophisticated tool guided by humans? The level of human involvement in AI content generation adds further complexity. Some AI outputs result from directly executing detailed instructions by a human operator. Others are the outcome of AI models that act more autonomously, drawing on vast datasets and generating surprising results beyond the original intent of their programmers. The nuances of this spectrum create a gray area regarding the degree of human authorship necessary for copyright claims. This lack of a clear legal definition directly impacts the ability to license, sell, or protect AI-generated content. If no one can definitively claim authorship, questions arise around who has the right to commercialize these creations, deal with potential infringement claims, and determine how these assets should be treated in the eyes of the law.

While a definitive, universal solution is still emerging, promising approaches exist to address the authorship and ownership issues surrounding AI-generated content. One approach involves revising existing intellectual property laws to explicitly address AI's role. This might include recognizing AI systems as creative tools, with ownership resting with the programmer or the person who directs the AI's output. Alternatively, new types of intellectual property rights specific to AI-generated works could be established.

Another potential solution lies in developing comprehensive contractual agreements and licensing models. These agreements would clearly outline the ownership rights and responsibilities of all parties involved in AI creation and content generation, including the developers, data providers, and users. This approach offers flexibility and allows for customized solutions tailored to specific use cases and industries.

CHAPTER 17   INTELLECTUAL PROPERTY RIGHTS WITH AI TECHNOLOGIES

Commercialization Concerns

Businesses investing in AI development want to protect their intellectual property. However, the collaborative and often open-source nature of AI innovation makes delineating clear ownership of algorithms, models, and the data they are built on difficult. Striking a balance between protecting investment and fostering innovation is essential. Knowing whether AI-generated content is crucial for ethical use and avoiding plagiarism concerns. Developing standardized labeling systems or metadata identifying AI contributions can help establish accountability and trust. Transparent disclosure about the use of AI in content creation helps avoid deception and builds confidence. Imagine reading a news article or viewing an image without knowing if it was written or created with significant AI involvement. Disclosing this information allows consumers to assess the work's potential biases and to make informed judgments about its reliability. Work is underway to develop standardized ways to indicate AI-generated or AI-assisted content. This could include visible watermarks, metadata embedded in files, or explicit labeling alongside the content. Having clear labeling systems empowers users to distinguish between human-created and AI-generated material, aiding them in evaluating its credibility and potential limitations.

Transparency in AI content generation helps address ethical concerns about manipulation and misinformation. Understanding the role of AI in producing content informs critical thinking and discourages the spread of potentially misleading information. It allows audiences to make their judgments rather than being unknowingly swayed by AI-generated outputs that may be biased or factually questionable.

CHAPTER 17   INTELLECTUAL PROPERTY RIGHTS WITH AI TECHNOLOGIES

Legal Frameworks and International Collaboration Can Help with AI IP Rights

Many legal experts advocate for updated copyright laws addressing AI authorship and ownership. These laws must balance the rights of human creators, AI developers, and the interests of those who use AI-generated outputs. Clear licensing agreements within the AI development landscape can clarify rights and responsibilities. These can address the use of training data, the ownership of derivative works, and the commercialization of AI-generated content. Technical solutions such as watermarking or blockchain-based registries can potentially track the provenance of AI content. This helps identify sources, manage ownership disputes, and establish authenticity. Since AI transcends borders, international cooperation is necessary to develop consistent legal frameworks and standards for ownership and sharing.

As the field rapidly develops, it is wise for businesses and individuals to use AI to adopt evolving best practices. These include transparency about AI use, careful sourcing of training data, implementing internal policies for responsible AI content generation, and staying informed about emerging regulations. Addressing these challenges requires ongoing dialogue among stakeholders: legal experts, policymakers, technologists, businesses, and content creators. Open collaboration and a shared understanding of the complexities are critical to finding solutions that protect innovation while ensuring ethical and fair use of AI technologies. Organizations like WIPO (World Intellectual Property Organization) are engaged in these discussions.

Corporate Attorneys can assess which aspects of AI models and systems qualify for intellectual property protection. This may include patents for novel algorithms or inventions embedded within the AI, trade secrets for proprietary training data or unique model architectures, and copyrights for generated code or elements with significant human involvement. Developing a comprehensive IP strategy helps secure the company's competitive advantage and deter unauthorized exploitation.

CHAPTER 17   INTELLECTUAL PROPERTY RIGHTS WITH AI TECHNOLOGIES

The legal and regulatory landscape surrounding AI is constantly evolving. Corporate attorneys must stay updated on new laws, emerging case law, and industry-specific AI guidelines. Proactive engagement in these discussions empowers attorneys to guide their organizations through this complex territory, ensuring compliance and minimizing risks associated with this groundbreaking technology.

Global Standards and Norms for AI Governance and Model Risk Management

Building a Global Framework for AI → Collaboration and Common Ground

The transformative power of Artificial Intelligence (AI) necessitates a global conversation to ensure its responsible development and deployment. Working with international bodies to establish shared standards and norms and fostering cross-border collaboration are key steps toward building a robust and ethical AI future. International cooperation is essential for developing a comprehensive set of standards and norms for AI. These standards would encompass algorithmic fairness, data privacy, human oversight, and transparency in AI decision-making. Existing international bodies like the Organization for Economic Cooperation and Development (OECD) or the UN Commission on International Trade Law (UNCITRAL) can facilitate discussions and foster consensus on these critical issues. Standardized principles would create a level playing field for AI development, encourage responsible innovation, and build public trust on a global scale.

Collaboration fosters a vibrant global AI ecosystem where knowledge exchange and joint research efforts can accelerate breakthroughs. International partnerships can leverage diverse perspectives and expertise, leading to more robust and comprehensive AI solutions for shared challenges. Imagine research consortiums tackling climate change with

CHAPTER 17   INTELLECTUAL PROPERTY RIGHTS WITH AI TECHNOLOGIES

AI-powered environmental modeling or international collaborations developing AI-driven healthcare solutions for global pandemics. By working together, nations can unlock the full potential of AI to address complex issues that transcend borders.

The rapid development of AI necessitates a unified approach. Collaboration with international bodies is crucial for establishing global standards and best practices for ethical AI development, deployment, and governance. This fosters trust, promotes responsible innovation, and mitigates potential risks associated with bias, privacy, and security concerns. International collaboration goes beyond setting standards. It unlocks the potential for knowledge sharing, joint research efforts, and coordinated approaches to tackling complex challenges like climate change or pandemics. Facilitating cross-border cooperation in AI can accelerate innovation, ensure the responsible use of AI technologies globally, and maximize the benefits of AI for all.

The Power of Shared Governance

International cooperation extends beyond technical development. Governance frameworks for AI necessitate a global approach. Joint efforts can create a unified approach to regulatory oversight, data governance, and risk mitigation strategies related to AI. Sharing best practices and establishing robust legal frameworks for AI across borders protects individuals and societies from potential misuse while fostering innovation within a well-defined ethical and legal landscape.

Transparency is vital in building trust in AI across borders. International collaboration can establish common ground for disclosing information related to AI systems, including how they function, the data they use, and the potential risks involved. Sharing this information allows for public scrutiny, fosters responsible development, and encourages international cooperation in addressing potential biases or unintended consequences inherent in some AI systems.

CHAPTER 17   INTELLECTUAL PROPERTY RIGHTS WITH AI TECHNOLOGIES

Summary and Thoughts

The Authorship Dilemma
A primary challenge revolves around the question of authorship and ownership. Current copyright laws in many jurisdictions primarily recognize human-created works. Even with some human input, AI-generated content falls into a legal gray area. This raises questions: does the AI tool creator have ownership? The person who provided the prompt? Can copyright even apply? Without clarity, it is difficult to determine who can rightfully control, license, or profit from AI-generated creations.

Detecting and Managing AI-Generated Content
The increasing sophistication of AI text, image, and code generators poses new challenges for sharing platforms. Distinguishing AI-generated content from human-created work is becoming complex, leading to concerns about plagiarism and the propagation of misinformation. Platforms must evolve reliable methods to track AI-generated content, manage attribution, and create policies to ensure ethical use and discourage abuse.

The Evolving Legal Landscape and Regulation
The issues surrounding AI-generated content and the underlying technologies demand legal guidance and potentially new regulations. Areas like copyright, intellectual property, liability for harmful content, and algorithmic bias require re-examination, considering AI's unique capabilities. Policymakers must balance protecting innovation and ensuring responsible AI use, fostering trust and accountability as AI becomes increasingly prevalent in creating and sharing content.

Global Collaboration for AI Governance
The international community can usher in an era of responsible and sustainable AI development by fostering collaboration and establishing common ground on AI standards, norms, and governance. This collaborative approach ensures that AI benefits humanity, promoting shared prosperity, tackling global challenges, and shaping a future where AI empowers individuals and societies worldwide.

CHAPTER 17   INTELLECTUAL PROPERTY RIGHTS WITH AI TECHNOLOGIES

Quiz

1. What is a core issue related to AI-generated content in terms of intellectual property?

   A) Identifying AI developers

   B) Determining ownership

   C) Enhancing algorithm efficiency

   D) Reducing computational costs

2. Which approach can help address the authorship and ownership issues surrounding AI-generated content?

   A) Ignoring copyright laws

   B) Recognizing AI systems as authors

   C) Revising existing intellectual property laws

   D) Minimizing human involvement

3. What complicates the delineation of ownership in AI innovation?

   A) Rapid development

   B) Open-source nature of AI

   C) High operational costs

   D) Limited collaboration

4. What can businesses develop to establish accountability and trust in AI-generated content?

    A) New AI algorithms

    B) Standardized labeling systems

    C) Proprietary datasets

    D) Minimal documentation

5. True or False: AI itself is recognized as an "author" in most copyright laws.

6. True or False: The collaborative nature of AI innovation makes clear ownership of algorithms and models difficult.

7. True or False: Developing comprehensive contractual agreements can help address the ownership rights and responsibilities of AI-generated content.

8. Businesses investing in AI development want to protect their intellectual property while fostering _____.

9. The lack of a clear legal definition directly impacts the ability to _____, sell, or protect AI-generated content.

10. Promising approaches to address AI authorship issues include revising existing laws or establishing new types of intellectual property rights specific to AI-generated works and developing comprehensive _____ agreements.

CHAPTER 18

Auditing AI Systems

Auditing AI systems is critical for ensuring these complex technologies operate ethically, reliably, and align with organizational goals. AI audits systematically examine AI models, the data they are built on, and their results. These analyses help identify potential biases, unintended consequences, technical flaws, and compliance risks. Regular audits promote fairness, transparency, and accountability in deploying AI – key factors in maintaining public trust and minimizing potential harm. When auditing AI systems, examine the AI governance framework, cybersecurity and privacy controls, data governance and protection (including its source, quality, and potential biases), and the model (including its algorithms, performance, and interpretability). I realize that this tall order of auditable items, however, is at a minimum and more importantly, ensure the following are prioritized:

Data → Scrutinize the training data used to build the AI model and the real-world data it processes during operation. Examine data sources for potential biases, inaccuracies, or incompleteness that could skew the AI's outputs. Assess data quality, privacy compliance, and whether the collection and use of data align with ethical standards and regulatory guidelines.

AI Model → Evaluate the AI model, including its architecture, algorithms, and decision-making processes. Look for any inherent biases in the model's design or training. Test the model's performance, accuracy, robustness, and interpretability. Identify any "black box" elements where the model's logic might be opaque, creating challenges for accountability and explainability.

CHAPTER 18   AUDITING AI SYSTEMS

Traditional IT Audit Programs vs. AI Audit Programs

Traditional IT Audit Programs evaluate existing information technology systems' controls, security, and operational efficiency. They typically assess risks related to data integrity, confidentiality, cybersecurity, system reliability, and compliance with regulations and internal policies. Traditional IT auditors often use well-defined checklists, sampling techniques, and process walkthroughs to identify vulnerabilities and recommend improvements.

The scope of traditional IT audit centers covers core IT components, operational processes (change management, access controls), and compliance with relevant frameworks (e.g., SOC 2, ISO 27001). The focus is on ensuring existing IT infrastructure's reliability, security, and efficiency. IT auditors typically employ a combination of techniques. They review documentation, interview IT staff, test system configurations, and analyze log data. Auditors often use well-established checklists and sampling techniques to identify potential control weaknesses or areas of non-compliance.

AI Audit Programs emerge as a specialized discipline to address the unique challenges AI systems pose. Unlike traditional software, AI systems are often dynamic, learning from data and potentially changing behavior over time. AI audit programs examine not only the technical infrastructure but also delve into the algorithms' complexities. Auditors assess issues like algorithmic bias, lack of transparency or explainability in AI decision-making, data quality concerns, and the potential for unintended consequences. Since AI auditing is young, standardized procedures and tools are still evolving.

The scope of AI audits goes beyond traditional IT concerns and encompasses AI systems' specific risks and complexities. This includes examining algorithms for bias and fairness, assessing training data quality, evaluating AI models' explainability, and ensuring ethical considerations are addressed throughout the AI lifecycle. AI audits require specialized expertise in machine learning and data science. They might involve statistical analysis of datasets, testing models with adversarial examples, and evaluating the potential societal impacts of AI deployments. Due to the evolving nature of AI, auditors must be adaptable and capable of critically assessing both technical and ethical implications.

AI Auditor Essential Skillsets

AI auditors must have comprehensive proficiency and experience in computer science, software engineering, and data science. They must understand AI concepts like machine learning algorithms, neural networks, data processing techniques, and the software development lifecycle, especially regarding AI systems. Understanding the specific industry and business context where AI is being deployed is crucial. AI auditors need to grasp the potential risks, regulations, and ethical considerations relevant to their auditing sector. For instance, auditing AI in healthcare requires expertise that is different from auditing AI in finance. AI auditing requires analytical skills, critical thinking, and an investigative approach. AI Auditors must identify patterns, anomalies, and potential vulnerabilities within complex AI systems. Traditional auditing skills like risk assessment, control evaluation, and evidence gathering remain essential and adapted to the unique challenges of AI.

AI Audit Program and Universe

Fairness, Data, and Explainability Audits. AI audits must rigorously examine algorithms for potential bias and discriminatory outcomes. This includes statistical analysis, testing for disparate impact, and identifying sources of bias in training data. A thorough evaluation of data quality, governance, and compliance with privacy regulations is paramount.

CHAPTER 18 AUDITING AI SYSTEMS

Data should be complete, accurate, and ethically sourced, with secure handling and clear lineage tracking. Assessing the explainability of AI models is essential, especially in high-stakes domains. Auditors probe the extent to which the model's reasoning can be understood and the trade-offs between explainability and performance.

Robustness, Security, and Ethics. AI systems must be robust against errors and deliberate attacks, such as adversarial examples designed to deceive them. Auditors assess security measures protecting AI models and their sensitive data. Contingency plans for unexpected outputs or malfunctions are also evaluated. The ethical implications of the AI system loom large in an AI audit. Alignment with ethical AI principles, proactive measures to ensure transparency and human oversight, and careful consideration of potential societal impacts are key areas of scrutiny for auditors.

Data Quality and Governance Audits. The quality of data used to train and power AI systems is paramount. AI audits evaluate data for completeness, accuracy, relevance, and potential biases. Additionally, auditors review data governance practices, including data lineage, data security, privacy safeguards, and compliance with regulations like GDPR.

AI Cybersecurity Audits. Auditing AI cybersecurity involves understanding these systems' distinct vulnerabilities beyond traditional IT security concerns. Adversarial attacks can target AI models that fool them into misclassification or manipulation. The data used to train AI models is also at risk, as poisoning data sources can corrupt the system. Additionally, AI models' complex, often opaque nature can make detecting security breaches or unauthorized access harder.

AI cybersecurity audits demand specialized techniques and expertise. Auditors assess security protocols surrounding the AI model, training data, and the underlying infrastructure. They test and validate controls for susceptibility to known adversarial attacks and evaluate encryption and access control measures. Auditors also examine incident response

CHAPTER 18   AUDITING AI SYSTEMS

plans specific to AI-related security breaches. They consider the need for ongoing monitoring of AI systems to detect anomalies and potential attacks early. Since AI cybersecurity is dynamic, auditors must stay updated on emerging threats and defensive techniques.

Bias Audits. Auditing for AI bias is a multi-stage process to identify and mitigate potential unfairness in AI decision-making. Auditors define the specific harms or discriminatory outcomes they investigate based on relevant regulations and ethical considerations. They then employ statistical metrics to measure potential bias across protected characteristics (e.g., race, gender) within the AI system's outputs. Testing with diverse datasets helps uncover disparate impacts on diverse groups. Auditors analyze the entire AI development lifecycle – from the selection and preprocessing of training data to the algorithm design and deployment context – to pinpoint the root causes of bias. Finally, they recommend mitigation strategies, such as rebalancing datasets, using bias-aware algorithms, incorporating human oversight, or conducting regular reviews for ongoing bias detection.

AI Governance Framework Audits. Auditing an AI governance framework involves a multifaceted examination to ensure it aligns with the organization's goals, industry standards, and ethical AI principles. This includes assessing the framework's comprehensiveness regarding algorithm development, data management, bias detection and mitigation, model explainability, accountability structures, risk assessment processes, incident response plans, and continuous monitoring systems. Auditors evaluate the existence and effectiveness of policies, procedures, roles and responsibilities, and the documentation process within the governance framework. They also consider the framework's flexibility to adapt to evolving regulations, technological advancements, and new societal concerns about how AI is used.

CHAPTER 18 AUDITING AI SYSTEMS

Auditing Process, Risk, and Controls

Identifying AI-specific Risks. Auditing AI process risks starts with thoroughly understanding the potential failure points and vulnerabilities unique to AI systems. These risks can stem from biased or incomplete training data, poorly designed algorithms, lack of explainability in decision-making, security breaches, unintended consequences due to AI model drift, or inadequate ethical oversight. Auditors must consider the critical processes throughout the AI lifecycle, including data collection, model development, deployment, and continuous monitoring.

Assessing Existing Controls. The next step is scrutinizing the controls implemented to mitigate these identified AI-specific risks. This includes safeguards for data quality and bias detection, processes for algorithm validation, explainability techniques, security measures protecting AI models and data, incident response plans for AI malfunctions, and ethical review mechanisms. Auditors evaluate the design and effectiveness of these controls, identifying gaps or areas where additional controls or stronger enforcement may be needed.

The risk profile and necessary controls vary depending on the application of the AI system. Auditing AI in high-stakes domains like healthcare or finance may demand more stringent controls and a higher level of explainability than less critical applications. Auditors tailor their assessment to the specific use case, evaluating the suitability of controls based on the potential severity of consequences if the AI system fails or produces biased outcomes. Develop an AI risk and controls matrix and map the risk landscape. The AI risk and controls matrix visually maps the identified risks and associated controls in a structured format. Rows typically represent the different risk categories, with columns listing specific risks within each category. Additional columns can include

- Likelihood and impact ratings for each risk.
- Existing controls are designed to mitigate the risks.

- Identification of any control gaps or areas where stronger controls are needed.

- Assignment of responsibility for implementing and monitoring controls.

- A process for periodic review and update of the matrix as the AI system or risk landscape evolves.

Why the Audit Committee Needs to Be Involved

Before the Audit Committee can effectively engage with AI audits, it must build AI literacy among its members. This may involve educational sessions on core AI concepts, the risks and benefits of AI for the organization, and the evolving regulatory landscape surrounding AI. Proactive education empowers the committee to ask informed questions, critically evaluate findings, and provide meaningful oversight.

The Audit Committee is responsible for overseeing organizational risk management and internal controls. As AI becomes increasingly integrated into core business processes, the risks associated with AI fall squarely within their purview. AI audits provide the Audit Committee with essential insights into the potential for algorithmic bias, security vulnerabilities, unexpected financial or reputational consequences, and the ethical implications of deploying AI systems.

Integrating AI into Audit Plans. AI audits should become a regular part of the Audit Committee's oversight plan. The committee must collaborate with internal auditors to define the scope, frequency, and focus of AI audits based on the organization's risk profile and the criticality of the AI systems in use. They should consider a phased approach, starting with high-risk AI applications and gradually expanding the scope of audits. The Audit Committee should establish clear communication channels between themselves, internal auditors, AI development teams, and management. Open dialogue is essential to ensure that audit findings are translated into actionable recommendations and that the organization is committed to responsible AI development and deployment. The Audit Committee can play a vital role in promoting a culture of transparency and accountability around AI within the organization.

Defining AI Audit and Risk Assessment Cycles

The ideal frequency of AI audits and risk assessments hinges on a risk-based approach. Organizations should start by identifying the AI systems used and the criticality of their functions. High-risk AI applications, such as those used in healthcare decision-making or financial trading, warrant more frequent and in-depth audits than low-risk systems like recommendation engines.

Factors Influencing Frequency. Several factors can influence the audit schedule:

- **Rate of Change** → Systems undergoing frequent updates or those operating in dynamic environments may need more regular audits to keep pace with evolving risks.

- **Regulatory Requirements** → Certain industries have specific regulations dictating minimum audit frequencies for AI systems.

- **Organizational Maturity** → Organizations new to AI deployment may benefit from more frequent initial audits to establish baselines and identify potential issues. As AI governance processes mature, audits can shift toward a less frequent but more targeted schedule.

Risk assessment should be an ongoing process. Even if a full-fledged AI audit is not performed frequently, organizations should monitor key risk indicators like bias metrics, data quality, system performance, and emerging security threats. This continuous monitoring allows for early detection of potential issues, informing the need for a deeper audit. While a risk-based approach is ideal, resource constraints may play a role in determining frequency. It is essential to balance comprehensive risk management, the availability of skilled auditors, and the allocated budget. Starting with risk prioritization and targeted audits of the most critical AI systems can ensure meaningful oversight even with limited resources.

Deciding between an independent AI auditor and an internal audit for AI audits is a complex choice, with pros and cons for each approach. Here is a breakdown of factors to consider:

Independent AI auditors often possess highly specialized expertise in auditing complex AI models, including technical skills in machine learning and deep domain knowledge of AI risks. This expertise might be challenging to build or maintain in-house, especially for smaller organizations.

External auditors can bring a fresh perspective and increased objectivity to the assessment. They are less likely to be influenced by internal biases or organizational pressures, which is essential for sensitive or critical AI applications. Independent auditors can offer valuable insights by benchmarking your AI systems and practices against industry standards and those of other comparable organizations.

A hybrid approach, where internal auditors collaborate with external specialists, can leverage the strengths of both models. This is especially beneficial when internal auditors need to upskill, when facing overly complex AI systems, or when dealing with sensitive domains where independent validation is valuable. The best choice depends on your organization's specific requirements, AI risk profile, budget, and the availability of both internal and external expertise.

CHAPTER 18   AUDITING AI SYSTEMS

Conducting an AI Readiness Assessment

Taking on the AI journey necessitates thoroughly assessing your organization's readiness. This initial step is imperative to ensure a business-aligned and successful integration of AI systems and technologies. The AI readiness assessment involves evaluating your existing infrastructure, cybersecurity countermeasures, data governance and management capabilities, talent pool, and overall organizational culture. By performing a SWOT analysis up-front, you can develop a tailored roadmap for AI adoption and governance, minimizing risks and realizing potential benefits.

Several frameworks can guide your AI readiness assessment. One popular option is the AI Readiness Index (AIRI), a model based on four pillars and nine dimensions, as illustrated in Figure 18-1.

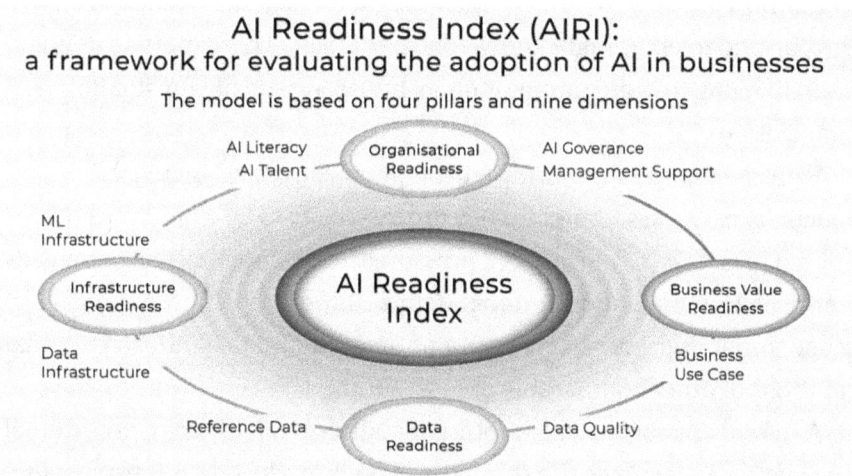

Figure 18-1. *AI Readiness Index (AIRI)*

(Grasso)

The AI Readiness Index evaluates a company's preparedness for incorporating Artificial Intelligence into business operations. The workshop-styled assessment criteria derive insights from collaborations with diverse industries and thought leaders. By identifying the discrepancy between current capabilities and desired goals, organizations can develop targeted strategies to optimize their AI implementation. This process transforms abstract ideas into actionable steps, thereby expediting the integration of AI within the company.

Assessing Organizational Readiness

The first phase of the readiness assessment evaluates the tone at the top about people, their knowledge and AI maturity levels, and their willingness to adopt AI systems across four dimensions.

Table 18-1. *Organizational Readiness Assessment*

Dimension	Assessment Performed	Outcome
AI Literacy	Evaluate the level of maturity and literacy regarding artificial intelligence, its use cases, and the current adoption of AI systems.	Document key findings and maturity levels.
AI Talent	Identify and evaluate whether the organization has internal resources to develop and manage AI models.	Document findings and AI proficient talent.
AI Governance	Identify and evaluate any defined strategic policies/framework to guide, oversee, and advise on the development and deployment of AI models.	Document findings and current state, whether complete or incomplete.
Tone-at-the-Top	Seek to understand and assess executive management's support of AI initiatives by allocating human and financial resources for its development within the organization.	Document findings and current support levels.

CHAPTER 18   AUDITING AI SYSTEMS

Organizational readiness evaluates an organization's capacity to undertake AI initiatives and systems. Essential components for successfully implementing advanced models and algorithms include a workforce possessing the required skill sets and a corporate culture fostering innovation.

Business Value Readiness Assessment

This assessment component solely focuses on identifying practical applications and real-world use cases. Key questions include: "Have you determined how AI can be implemented in your business?" and "What is the potential return on investment for your organization by incorporating AI?"

Table 18-2. Business Value Readiness

Dimension	Assessment Performed	Outcome
Business Value	Evaluate and assess whether the organization has identified business processes or lines of businesses that can benefit from AI systems and create business value or competitive advantages.	Document results and findings.

Strategic planning that outlines innovation areas and pinpoints specific use cases can streamline the adoption of complex technologies such as AI. Organizations can strategically select use cases where AI implementation will yield the highest efficiency by determining the value proposition and the potential value generated.

Data Readiness Assessment

This element, encompassing two dimensions – data Quality and Reference Data – aims to assess an organization's preparedness to manage the data that AI models and algorithms will utilize.

Table 18-3. Data Readiness Assessment

Dimension	Assessment Performed	Outcome
Data Quality	Evaluate and assess the organization's data governance and management processes to ensure the accuracy and completeness of the data it collects to fuel AI models.	Document data governance processes and control standards, along with key findings.
Reference Data	Evaluates both the quality and accessibility of data to determine an organization's preparedness to utilize AI models and algorithms effectively	Document findings, i.e., varying levels of data quality and reference data preparedness, necessitating targeted improvements to ensure optimal utilization for AI initiatives.

As mentioned throughout this book, data acquisition and quality are paramount, fueling AI model training. In contrast to traditional software development, where programmers meticulously dictate operations, AI models learn from experience, akin to human learning, through training datasets. As a result, the importance of data governance has escalated, ensuring the quality, accuracy, and unambiguity of these datasets.

In AI model training, data standardization is essential to prevent misinterpretations when different departments use the same data entities in varying formats. Establishing a centralized data repository ensures consistent referencing and interpretation of data across the organization, thus facilitating seamless collaboration and enhancing the accuracy and effectiveness of AI models.

Infrastructure Readiness Assessment

The infrastructure readiness assessment provides insight into the organization's ability to support AI systems and models. Whether or not your IT environment can sufficiently support your AI initiatives lies in two dimensions – data infrastructure and machine learning infrastructure.

Data infrastructure encompasses a range of components: servers for processing, storage for archiving, and the organizational processes, policies, and guidelines that govern how data is managed. These latter components, collectively known as Data Governance, play a critical role in establishing a formal framework for managing data throughout its lifecycle.

A machine learning infrastructure encompasses the personnel, organizational workflows, and technology required to build, train, and deploy machine learning models. Often referred to as AI infrastructure or a constituent of MLOps (Machine Learning Operations), this infrastructure facilitates the reliable and efficient deployment and maintenance of machine learning models in production environments.

Table 18-4. Infrastructure Readiness Assessment

Dimension	Assessment Performed	Outcome
Data Infrastructure	Evaluate and assess the organization's data infrastructure and governance models to ensure that data is inventoried, managed effectively, securely, and complies with regulations.	Document data governance processes and control standards, along with key findings.
Machine Learning Infrastructure	Evaluate and assess the machine learning infrastructure components and processes that support the entire lifecycle of machine learning models.	Document findings, i.e., ML landscape and development environment(s).

The Final Score and Evaluation

The AIRI system employs a comprehensive assessment methodology to evaluate an organization's AI readiness. This process involves analyzing nine distinct dimensions across four key categories, culminating in a calculated score that reflects the organization's overall preparedness for

AI implementation. The resulting score is then mapped onto a matrix, providing a definitive assessment of the organization's current AI readiness level, like in Figure 18-2.

	AI Unaware	AI Aware	AI Ready	AI Competent
Average Score	Less than 1	1 to 1.9	2 to 2.5	More than 2.5
General Capabilities	Might hear about AI but is unaware of applications.	Savvy consumers of AI solutions. Capable of identifying use cases for AI applications.	Capable of integrating pre-trained AI model into products or business processes.	Capable of developing customised AI solutions for specific business needs.
General Characteristics	Wait for vendors to convince use cases and business value of AI.	Identified potential use cases and seek AI solutions from vendors.	Evaluated viability of pre-trained AI models.	Developed roadmap for AI implementation.
AI Adoption Suitability	Consume ready-made, end-to-end AI solutions.		Integrate pre-trained AI models and solutions for common AI applications.	Developed customised AI model for unique business needs.

Figure 18-2. AIRI Final Score and Readiness Matrix

(Grasso)

Other frameworks, such as the Gartner AI Maturity Model, focus on evaluating your organization's progress along the AI maturity curve. Regardless of your framework, the assessment should provide a comprehensive overview of your current state and highlight areas that require improvement before embarking on your AI journey.

Addressing Exceptions, Findings, and Remediation

Addressing AI audit findings and exceptions is crucial in building robust and responsible AI systems. There are several approaches to tackle this effectively. Begin by triaging the audit findings. Not all exceptions carry equal weight. Focus on critical findings with a high potential for monetary loss, reputational damage, discriminatory outcomes, or significant security risks. Prioritize these for immediate remediation while developing plans

CHAPTER 18   AUDITING AI SYSTEMS

to address lower-risk issues promptly. Go beyond merely addressing the symptoms of the identified exceptions. Thoroughly investigate the root causes within the AI system's design, the data used for training, or the surrounding processes. This understanding is key to preventing similar issues from recurring in the future. Was the exception caused by inadequate data quality, a flawed algorithm design, insufficient bias testing, a lack of explainability safeguards, or a combination of factors? Develop targeted remediation plans addressing the root causes. These may include

- Retraining AI models with more balanced or representative datasets

- Implementing bias mitigation techniques or incorporating explainability tools

- Enhancing data governance and quality control processes

- Strengthening cybersecurity measures for AI systems and their data

- Revising ethical review procedures and incorporating fairness metrics into monitoring

Accountability, Communication, and Continuous Improvement. Establish clear ownership of remediation actions, timelines, and metrics to track progress. Maintain open communication between auditors, AI development teams, and management throughout the process. Treat AI audit findings as opportunities for improvement. Incorporate lessons learned into your AI governance framework, refining processes to minimize risks and build more robust, trustworthy AI systems in the future.

CHAPTER 18   AUDITING AI SYSTEMS

Specific Remediation Strategies

Addressing Algorithmic Bias → If bias is detected, consider techniques like rebalancing training datasets to reduce the underrepresentation of certain groups, employing bias-aware algorithms designed to mitigate unfairness, or incorporating human-in-the-loop processes for critical decisions to provide oversight. If bias stems from imbalances in your training data, focus on collecting additional data representing underrepresented groups or use data augmentation techniques to create synthetic samples. Be cautious of the potential to introduce new biases even with these approaches. Explore algorithms specifically designed to mitigate bias. These include techniques like counterfactual fairness (altering inputs to see if model outputs change unfairly), preprocessing data to remove correlations with protected attributes, or post-processing model outputs to adjust for disparities.

Human-in-the-Loop → Incorporate human oversight for high-stakes decisions. A human expert can review AI outputs, especially those flagged by bias-detection tools. This adds a layer of scrutiny and ethical judgment.

Improving Data Quality and Governance → Establish stricter data quality checks and implement processes to identify and rectify errors or inconsistencies in data sources. Enhance data privacy controls, adhering to regulations and ethical data collection and handling principles. Formalize data lineage tracking to understand data origins and transformations. Implement rigorous data validation and cleaning procedures – detecting and correcting outliers, handling missing values, and ensuring consistency – document data preprocessing steps to increase transparency and reproducibility of results. Implement robust data lineage tracking systems that capture how data is collected, transformed, and used throughout its lifecycle. This aids in understanding the origins of bias and auditing for compliance and facilitates easier debugging in case of issues.

Enhancing Explainability → If the AI model lacks interpretability, explore techniques like feature importance analysis, local explanations (explaining individual predictions), or using inherently more explainable model types for high-stakes domains. These help understand how the AI system makes decisions, aiding in debugging and promoting trust. Use techniques to identify the input features the AI model relies on most heavily for its decisions (e.g., permutation importance, SHAP values). This can highlight potential biases if the model focuses on irrelevant features correlated with protected attributes. Tools like LIME or SHAP explain individual predictions, showing how changes in input features impact the output. This aids in understanding specific decisions and identifying potentially problematic cases. In high-stakes domains, consider using model classes like linear models or decision trees, which are easier to interpret by design. Always consider the trade-off between explainability and potential performance sacrifices.

Remediation is often an iterative process. Regular monitoring and follow-up audits are essential to ensure that implemented changes effectively address the issues and mitigate risk.

Corrective Action Plans (CAP)

Clarity and Specificity → CAPs should clearly outline the actions to be taken, the responsible parties, completion timelines, and metrics to track progress. Avoid vague or overly broad action plans.

Collaboration and Accountability → Involve stakeholders from the AI development team, data scientists, IT security, and potentially compliance or legal teams in developing and implementing the CAP. Clear ownership for each corrective action is paramount.

Documentation and Feedback Loop → Document the CAPs, ensuring version control for tracking changes and measuring the effectiveness of implemented actions. Utilize findings from post-remediation audits further to refine AI development, deployment, and governance processes, fostering continuous improvement.

Summary and Thoughts

AI auditing is a specialized discipline that scrutinizes the unique risks associated with Artificial Intelligence systems. These audits go beyond traditional IT audits, evaluating algorithm fairness, data quality, model explainability, cybersecurity, and the potential for unintended consequences. Auditors employ a mix of technical assessments, statistical analysis, and domain-specific expertise to identify potential vulnerabilities within AI systems.

AI audit findings are crucial for building trust and ensuring responsible AI development. These findings may expose biases in AI models, inadequacies in data governance practices, security loopholes, or a lack of transparency in decision-making. Remediation strategies focus on root cause analysis and tailor corrective actions. This may involve rebalancing datasets to mitigate bias, enhancing data quality checks, implementing explainability tools, or tightening security around AI systems. Clear corrective action plans outlining specific steps, timelines, and responsible parties are essential to guide the remediation process.

The Audit Committee plays a vital oversight role in AI risk management. Proactive education on AI concepts is key to empowering the committee to engage effectively with AI audits. Integrating AI audits into the committee's oversight plan ensures regular assessment of AI-related risks within the organization. Open communication between the Audit Committee, internal auditors, AI teams, and management fosters a culture of transparency, accountability, and continuous improvement in AI development practices.

CHAPTER 18   AUDITING AI SYSTEMS

Quiz

1. What is a primary focus of auditing AI systems?

    A) Reducing development time

    B) Ensuring ethical and reliable operation

    C) Minimizing computational resources

    D) Enhancing user experience

2. Which element is NOT typically examined in AI audits?

    A) Data quality

    B) Model performance

    C) Marketing strategies

    D) Cybersecurity controls

3. What is a key difference between traditional IT audit programs and AI audit programs?

    A) AI audit programs focus on algorithm transparency

    B) Traditional IT audits prioritize cybersecurity only

    C) AI audit programs ignore data sources

    D) Traditional IT audits are less thorough

4. What should AI audit programs prioritize to maintain public trust?

    A) Minimizing costs

    B) Ensuring fairness and accountability

    C) Reducing data usage

    D) Ignoring stakeholder feedback

CHAPTER 18 AUDITING AI SYSTEMS

5. True or False: Regular audits promote transparency and accountability in AI deployment.

6. True or False: AI audits do not need to examine the data sources used for training models.

7. True or False: Evaluating the AI model's design and decision-making processes is essential in AI audits.

8. Auditing AI systems helps identify potential biases, unintended consequences, technical flaws, and _____ risks.

9. Regular audits promote fairness, transparency, and _____ in deploying AI.

10. AI audit programs should scrutinize the _____ data used to build and operate the AI model.

CHAPTER 19

AI Model Inventory and Facts

As organizations increasingly integrate AI into their core processes, it becomes imperative to have a robust understanding of the AI models driving these systems. This is where an AI Model Inventory and accompanying AI Factsheets come into play, offering essential tools for AI governance, transparency, and responsible deployment.

An AI Model Inventory is a centralized repository documenting key information about each AI model used across an organization. It goes beyond simply listing models and probes into their purpose, technical details, training data characteristics, performance metrics, known limitations, and potential risks. Think of it as the "passport" of an AI model, providing a comprehensive overview of its background and capabilities. AI Factsheets complement the Model Inventory by offering a standardized format to communicate essential model information clearly and concisely. These factsheets aim to bridge the gap between technical experts and stakeholders, promoting transparency and enabling informed decision-making. They typically cover model performance, intended use cases, ethical considerations, and known limitations or biases.

Establishing a well-maintained AI model inventory and utilizing AI factsheets can help organizations foster greater accountability, manage risks proactively, and ensure AI systems align with ethical guidelines, regulatory frameworks, and overall business objectives. I will summarize and articulate the AI model inventorying process and discuss the key strategy points.

Model Identification and Factsheet

The first step is to identify all AI models used across the organization. This may involve decentralized models used by various teams or legacy systems. Strategies include conducting surveys or interviews with relevant departments, leveraging automated discovery tools to scan systems or source code repositories, and establishing a centralized reporting process for new AI model development. Once models are identified, comprehensive information will be collected for each. This may include

- **Technical Details** → Algorithm type, architecture, programming language, dependencies
- **Data** → Sources of training data, data volume, preprocessing techniques
- **Purpose and Use case** → Business problem the model solves, how outputs are used
- **Performance Metrics** → Accuracy, precision, recall, and other domain-specific metrics.
- **Risks and Limitations** → Known biases, security vulnerabilities, explainability limitations
- **Ownership and Maintenance** → Teams responsible for development, updates, monitoring

Common Challenges
Choose a suitable format for the AI Model Inventory (spreadsheets, databases, or specialized software). Structure information logically to allow for easy filtering, searching, and auditing. Crucially, the inventory is not

a static document. Implement a process for regular updates as AI models are created, modified, or retired. Assign clear ownership to maintain the inventory, ensuring its accuracy and ongoing usefulness. Organizations often have decentralized AI development with teams that build and deploy models independently. This lack of central oversight makes it difficult to track all models in use and ensure consistent information capture. Employees may use off-the-shelf AI tools or create simple models without involving the central IT or governance teams. These "shadow AI" models can easily fall under the radar. AI is a rapidly evolving field. Models are frequently updated, new ones are introduced, and some become obsolete. Keeping the inventory updated with this constant churn poses a significant challenge. Without standardized processes for documenting and reporting AI models, the inventory may suffer from incomplete, inconsistent, or outdated information. This undermines its usefulness for governance and risk management. Some teams may perceive inventorying as added bureaucracy, hindering adoption and compliance.

Strategies to Address Challenges
Establish a centralized AI governance function or task force to oversee the inventorying process. This team sets standards, defines reporting mechanisms, and coordinates across departments. To identify AI models, utilize tools that automatically scan systems, code repositories, and network traffic. These tools can reduce the manual effort required for discovery, especially in large, complex organizations. Create incentives for teams to report new AI model development proactively. This encourages collaboration rather than hindering innovation.

Initially, avoid overly complex documentation requirements. Focus on the most critical information to capture, streamlining the process and fostering buy-in. Educate teams across the organization about the importance of the AI Model Inventory and the potential risks of mismanaged AI. This helps shift the culture toward a proactive and collaborative approach to AI governance.

Types of Automated Discovery Tools

Code and infrastructure scanning tools analyze source code repositories, development environments, and containerized applications (e.g., Docker images).

They aim to identify

Libraries and frameworks used (TensorFlow, PyTorch, etc.) indicate the presence of AI models. Specific model objects within code, extracting model parameters or file names. API calls to external AI services for cases where models are not hosted internally. Tools that monitor network traffic can detect patterns that reveal interactions with AI models.

This is useful for identifying AI services utilized as cloud-based solutions or APIs where the code may not be accessible within the organization. Some file formats associated with AI models contain metadata (e.g., ONNX, TensorFlow SavedModel). Specialized tools can scan filesystems or data stores, extracting model architecture, hyperparameters, and potentially some aspects of training data characteristics. Many commercial AI governance platforms offer discovery solutions that combine the above techniques for a more comprehensive approach.

Considerations When Choosing a Discovery Tool

Select a tool that aligns with your organization's infrastructure and development practices. A tool focused on code scanning might be less effective if your models are primarily deployed as external services. Evaluate the level of detail the tool can extract. Look for tools to integrate with your existing inventory system (database, spreadsheet, specialized software) for streamlined workflows. Assess the tool's accuracy in correctly identifying AI models and minimizing false positives to avoid excessive manual verification.

Automated discovery tools are powerful aids but should not be relied on entirely. They typically work best with well-defined reporting processes and a strong emphasis on organizational communication. Choosing a discovery tool is an iterative process. Start by prioritizing your most pressing needs, experiment with different options, and gather feedback from internal stakeholders to refine your selection over time.

The following are questions you must answer when selecting a discovery tool.

Alignment with Technical Environment

Deployment Models. How are your AI models deployed? Are they embedded in applications, running as containerized services, or primarily accessed via cloud-based APIs? Choose a tool that can effectively scan your dominant deployment environment.

Programming Languages and Frameworks. What programming languages (Python, Java, etc.) and AI frameworks (TensorFlow, PyTorch, Scikit-learn) are your teams using? The tool should be able to analyze the code and libraries commonly used within your organization.

Data Storage. Where is the data used to train and run your models stored? If your data resides in a specific database or file storage format, consider tools to analyze metadata from those sources.

Depth of Discovery

Basic Identification → Do you need to identify the presence of AI components within systems? Or do you require deeper analysis?

Model Parameters → Do you need the tool to extract model architecture (number of layers, neuron types), hyperparameters, or training dataset characteristics? This level of detail might be necessary for risk assessment or regulatory compliance.

Integration Capabilities

Inventory System → How will the discovery tool's output feed into your existing AI Model Inventory? Look for tools with APIs, data export formats, or pre-built integrations that align with your chosen inventory solution.

Monitoring and Alerting → Can the tool integrate with your monitoring systems to provide real-time alerts about new or potentially unauthorized AI model deployments? This facilitates proactive governance.

Accuracy and False Positives

Precision vs. Recall → Consider whether you are more concerned about missing AI models (the tool has low recall) or about the tool identifying too many non-AI components as potential models (the tool has low precision). This trade-off depends on your risk tolerance.

Manual Verification → Automated discovery will require manual verification by human experts. To streamline the review process, assess the tool's ability to prioritize or flag potentially problematic models.

Additional Factors

Scalability → Will the tool function effectively as your organization's AI footprint grows and your IT environment becomes more complex?

Cost → Consider the pricing model (license fee, per model, per scan) and align this with your budget and projected scale of AI model usage.

Usability → Is the tool user-friendly for technical teams and those involved in the AI governance process? You must ensure your chosen solution is the most user-friendly and easy to deploy.

Vendor Support → Evaluate the vendor's reputation, documentation, and responsiveness in case you need technical assistance with the tool's deployment or tuning.

Summary and Thoughts

An AI Model Inventory is a centralized repository that houses vital information about each AI model employed within an organization. This inventory is akin to a detailed passport for every AI model, providing a comprehensive overview of its purpose, technical specifications, training data, performance metrics, limitations, and potential risks. A well-

CHAPTER 19   AI MODEL INVENTORY AND FACTS

structured inventory fosters transparency, enables effective AI governance, and helps ensure the responsible use of AI across the enterprise.

AI Factsheets complement the Model Inventory by presenting model information in a standardized, easily digestible format. They are intended for technical and non-technical stakeholders and summarize key aspects like model performance, intended use cases, known biases or limitations, and relevant ethical considerations. By promoting a clear understanding of the AI systems, factsheets empower informed decision-making and facilitate ongoing monitoring aligned with an organization's goals and AI principles.

Quiz

1. What is the primary purpose of maintaining an AI Model Inventory?

    A) Reducing operational costs

    B) Enhancing transparency and management

    C) Limiting model usage

    D) Increasing data redundancy

2. Which aspect is crucial when creating an AI factsheet?

    A) Ignoring data sources

    B) Providing detailed model documentation

    C) Minimizing stakeholder input

    D) Reducing transparency

CHAPTER 19 AI MODEL INVENTORY AND FACTS

3. What challenge(s) does maintaining an AI Model Inventory address?

    A) Increased development costs

    B) Model identification and tracking

    C) Reducing computational resources

    D) Enhancing user experience

4. Why is it important to keep the AI Model Inventory updated?

    A) To increase operational complexity

    B) To ensure accuracy and relevance

    C) To limit model usage

    D) To reduce documentation efforts

5. True or False: An AI Model Inventory helps in managing and governing AI models effectively.

6. True or False: Detailed model documentation is not necessary for AI factsheets.

7. True or False: Keeping the AI Model Inventory updated is crucial for maintaining its accuracy and relevance.

8. An AI Model Inventory provides a comprehensive overview of all AI models, enhancing _____ and governance.

9. AI factsheets should include detailed documentation on model development, _____, and usage.

10. Maintaining an updated AI Model Inventory ensures models are accurately identified and _____.

CHAPTER 20

AI and Enterprise Architecture

The Strategic Union: Unveiling the Empowering Role of Enterprise Architecture in AI Initiatives

AI affects businesses in many sectors. Still, organizations must build a solid base and a clear way to move forward to get the most out of their AI projects. That's where enterprise architecture (EA) comes in. EA connects high-level AI goals to implementation resources and ensures that companies' enterprise don't waste money on projects that might look good but ultimately produce little value. In this brief chapter, we'll see what EA looks like in practice. From conception to development to ongoing management, EA can and should play a role that keeps AI viable and valuable well into the future.

Enterprise Architecture holds the potential to bolster AI systems in organizations substantially. Primarily, it allows for a full view of a business and its technology, processes, and data flows. AI is often sold as something that will supercharge all aspects of a business and its processes. The issue, of course, is that coping with the intoxicating value proposition of AI leads leaders to make decisions in a way that's often completely divorced from the existing structure and state of the business. Enterprise Architecture, by contrast, is about knowing everything happening everywhere and understanding how all the components work together. EA smooths the fusion of AI into the company's environment. It forges the information flow and infrastructure necessary for AI systems to function effectively. It enables the procedures that manage the often-elusive aspects of AI ethics and governance, including a clear pathway for assuring that AI systems are fair, that they can be understood and explained, that they are accountable, and that by using AI, an organization is not doing something that violates the public trust. This iterative approach allows for continuous improvement and a desirable amalgamation of EA and AI, ensuring their long-term success and value generation within the organization.

Moreover, EA's most important function when it comes to the business of artificial intelligence is scaling. When an AI model proves successful in one part of the business, it can also be used in other parts. But that won't happen unless the use and performance of that first instance are overseen by a team with an eagle eye for the enterprise IT picture. EA is the lubricant that eases the flow between an organization's use of AI in different places and the profound visibility one gets into the consequences of that use in several instances – or, in the worst cases, when an AI model has issues in one place that must be cleaned up in another.

The Amalgamation of AI, EA, and Business Strategies

Before diving into complex algorithms, a clear understanding of business goals is paramount. The convergence of Artificial Intelligence, Enterprise Architecture, and Business Strategies signifies a transformative shift in the corporate landscape, especially concerning AI governance and risk management. With its ability to analyze vast datasets and generate

CHAPTER 20 AI AND ENTERPRISE ARCHITECTURE

insights, AI is becoming integral to decision-making processes. As the blueprint of an organization's IT infrastructure, EA provides the framework for integrating AI solutions, ensuring they align with overall business goals. When business strategies are formulated with a keen awareness of AI's potential and EA's role in its deployment, organizations can unlock new avenues for growth and innovation. However, this amalgamation raises critical questions about AI governance and risk management.

Regurgitating the definition of AI governance and risk management, AI governance is the system of rules, practices, and processes that ensure AI is developed and used responsibly. It encompasses ethical considerations, regulatory compliance, and establishing accountability frameworks. Risk management in AI involves identifying and mitigating potential harms, such as bias in algorithms or misusing AI-generated data. Integrating AI, EA, and business strategies necessitates a robust governance model that addresses these risks while maximizing AI's potential benefits. EA can be pivotal in operationalizing AI governance by ensuring that AI systems are designed and implemented according to the organization's governance model and risk frameworks.

By providing a holistic view of the organization's technological landscape, EA can help identify and mitigate risks while strengthening an organization's governance principles. Organizations can create a culture of responsible AI use by integrating AI governance principles into EA frameworks. AI is deployed to create value while safeguarding against potential harm and allowing organizations to harness AI's power while ensuring ethical and responsible use. This integrated approach mitigates risks and fosters a culture of trust and transparency, which is essential for the long-term success of AI initiatives.

EA plays a vital role in this union by

- Collaborating with business stakeholders, EA can identify areas where AI can automate tasks, improve decision-making, or enhance customer experiences.

- Working with AI actors to establish clear metrics (as mentioned in this book) for AI systems and ensure alignment with the goals and objectives of the business. EA can also aid in defining how AI metrics or KPIs will be measured, whether it's quantitative savings, increased efficiency, or improved customer satisfaction.

Designing the AI Landscape with Enterprise Architecture

At the broadest level, the role of enterprise architecture in AI is to enable a clear understanding and effective management of the organization's AI landscape. A traditional role of enterprise architecture has provided a framework for identifying and integrating components comprising the organization's largely digital world. This is not unlike what is needed for AI, where a dizzying array of software platforms, interface technologies, and data formats must be understood, coordinated, and managed. Presumably, working from within the structure of an enterprise architecture allows an organization to see the potential for synergy among AI systems, identify points of significant risk, and use (or not use) the right AI tools for AI systems' management and administration.

Once the business case is established, EA can design the architectural blueprint for implementing AI systems. The blueprint activities include but are not limited to the following:

- Assessing the existing IT landscape and infrastructure. EA can identify gaps or potential compatibility issues with the chosen AI systems and help select the appropriate technology stack, considering factors like scalability, security, and integration capabilities.

- EA can help design a data architecture that facilitates seamless data flow across various systems and ensures data quality for model training and performance.

When governing artificial intelligence and managing its risks, EA can help establish standards and governance principles to ensure consistency and control in AI development and deployment. This includes guidelines for data access, model development practices, and ethical considerations. Organizations can develop comprehensive processes for AI governance, model risk management, and the like. Still, if those aren't well integrated within the organization's EA strategies – i.e., conflicting AI frameworks and structures and unclear lines of authority – they become much less likely to achieve the desired results. By aligning with and integrating EA frameworks, AI governance efforts – essentially, efforts to ensure that the organization's AI capabilities are developed and used in an accountable, transparent, and not overly risky manner – can be vastly strengthened.

Enterprise Architecture during AI Implementation and Ongoing Management

AI implementation and management can be complicated, and contrary to popular belief, they require a well-aligned strategy. One of the best strategic tools for businesses working with AI is the alignment with enterprise architecture. From an AI implementation standpoint, the primary purpose of EA is to provide a comprehensive overview of the organization's current technological landscape and the different components that make up that landscape. By doing so, EA can serve as a beacon for the organization – leading it from where it is now, in terms of technology, to where it wants and needs to be in the future, all while identifying the kind of AI systems that'll work within its infrastructure.

CHAPTER 20   AI AND ENTERPRISE ARCHITECTURE

Enterprise architecture reduces technology integration or incompatibility risk during the AI deployment phase. By identifying and analyzing potential issues or vulnerabilities, EA can quickly and proactively mitigate them. Besides addressing the IT aspects of risk, EA also considers "model bias" and regulatory issues that can affect a company's image when using AI systems.

EA's work doesn't stop at implementation but extends into the ongoing management of AI systems. It can help establish mechanisms for continuous evaluation and ensure that the AI models reliably and accurately produce the desired business outcomes. AI keeps improving and getting "smarter"; however, if left unchecked, it could evolve and behave in ways that stray from the organization's intent and strategic goals.

Here are a few prescriptive roles your EA can play during AI implementation and ongoing management and administration.

- EA can facilitate the integration and interoperability of AI models with existing enterprise systems, ensuring smooth data flow and operation.

- EA can design the AI systems architecture to be scalable and future-proof, allowing the AI solution to adapt to evolving business needs and business process automation requirements.

- EA can work with IT, business process areas, cybersecurity, AI governance, and model risks teams to monitor AI model performance in production and identify potential issues or areas for improvement.

Failing to involve Enterprise Architecture (EA) in the AI implementation and ongoing management phases can have far-reaching and detrimental consequences and possesses what I call the "Frankenstein model" – namely, where different AI systems have been patched together but fail to integrate effectively with the existing infrastructure and ecosystems of the organization. When EA is not involved, you risk technological misalignment and situations where AI initiatives are not in sync with the broader business objectives. This misalignment results in inefficient AI systems that do not align with business automation objectives, wasted capital expenditures, discouraged teams that work on the AI systems, and inaccurate AI models.

The Benefits of Amalgamating AI, EA, and Business Strategy Teams

When AI, EA, and business strategy teams work together, a powerful synergy pushes organizations toward greater innovation and efficiency. Leading organizations reap the benefits of AI, harnessing it to carry out tasks in ways we previously reserved for human intellect. By creating better networks across disciplines, you not only identify opportunities for the way AI can drive change but also for the ways that change can be made. ... It's a system that, when working properly, aligns the technology of AI with the vision of where the organization wants to go strategically. It allows the way change is enacted to be done in a manner that creates business value. Through collaboration, they can determine useful sources of intelligent automation, implement good rules for governing what can and can't be done with it, and ensure the quality of AI systems and their training data.

When AI, EA, and business strategy teams integrate, they form a dynamic ecosystem where innovation reigns, risks are kept at bay, and top-line value is maximized. Collaboration between these three groups guarantees the top-down effect needed to make AI initiatives successful and ensure they are part of the long-term corporate plan. Further, it is a way for organizations to stake their claim in the AI future and become the epitome of good AI governance and model risk management.

CHAPTER 20   AI AND ENTERPRISE ARCHITECTURE

Summary and Thoughts

Bridging the gap between artificial intelligence technology and business objectives is necessary throughout the AI development lifecycle, and it's also where Enterprise Architecture proves most useful. EA shouldn't just be the team that draws fancy technical diagrams and charts. Rather, they are the central intelligence for an organization's technological landscape. That's a fancy way of saying they identify, evaluate, and define how to integrate the best AI systems conducive to achieving your organization's strategic goals. EA provides AI actors and stakeholders with a perspective on "the big-tech picture." They take the current state of your organization's technology environment and infrastructure and ensure the lines of business processes are efficiently interoperating. For an enterprise to adopt artificial intelligence, it must understand the essential cooperation between EA and AI. The natural combination of these two disciplines allows for the range of necessary prerequisites to be put in place. With a solid understanding of the what, why, where, and how of integrating AI and the EA framework, followed by putting that integrated EA framework into action, an organization can achieve the adoption of AI and harness its power far more effectively in far shorter time frames. Indeed, it is the combination of EA and AI within an integrated framework that allows for the necessary people, processes, and technology to deliver the real power of AI to the organization.

CHAPTER 20   AI AND ENTERPRISE ARCHITECTURE

Quiz

1. What is the strategic role of enterprise architecture in AI initiatives?

    A) Reducing AI deployment time

    B) Providing a holistic view of the organization's technological landscape

    C) Limiting AI innovation

    D) Increasing data redundancy

2. Which benefit does the integration of AI and enterprise architecture NOT offer?

    A) Ensuring scalability

    B) Addressing governance and risk management concerns

    C) Reducing operational efficiency

    D) Facilitating technological integration

3. How does enterprise architecture support AI implementation?

    A) By ignoring stakeholder input

    B) By facilitating technological integration and scalability

    C) By reducing documentation efforts

    D) By minimizing transparency

CHAPTER 20   AI AND ENTERPRISE ARCHITECTURE

4. What is a key focus area when designing the AI landscape with enterprise architecture?

   A) Reducing computational resources

   B) Enhancing technological integration

   C) Minimizing stakeholder involvement

   D) Limiting model usage

5. True or False: Enterprise architecture provides a holistic view that supports AI initiatives by addressing governance and risk management concerns.

6. True or False: Integrating AI with enterprise architecture does not enhance operational efficiency.

7. True or False: Enterprise architecture plays a crucial role in ensuring the scalability of AI initiatives.

8. Enterprise architecture empowers AI initiatives by providing a holistic view of the organization's _____ landscape.

9. Integration of AI and enterprise architecture enhances _____ and facilitates technological integration.

10. When designing the AI landscape, a key focus area is enhancing technological _____.

CHAPTER 21

Aligning AI Governance, AI Development Lifecycle, and Systems Development Lifecycle Processes

The fusion of AI Governance, AI Development Life cycle (AIDLC), and System Development Life cycle (SDLC) processes share a common foundation of prioritizing responsible and ethical AI development. Both emphasize the need for clearly defined guidelines and principles that guide the entire lifecycle of AI systems. This includes establishing accountability structures, risk assessment procedures, and clear documentation practices. Both frameworks also stress the importance of data ethics, ensuring that the data used to train and operate AI models is collected and handled responsibly, focusing on fairness, privacy, and transparency.

CHAPTER 21   ALIGNING AI GOVERNANCE, AI DEVELOPMENT LIFECYCLE, AND SYSTEMS
              DEVELOPMENT LIFECYCLE PROCESSES

Aligning the trinity helps build ethical, secure, and reliable AI systems. This integrated approach weaves in checks and balances at every stage, ensuring responsible AI development, validation, approvals, deployment, and stage gates. It minimizes risks and maximizes the positive impact of AI on society. By embedding the AI development processes into each feasible stage of the SDLC, potential issues of incompatibility, security vulnerabilities, and unintended consequences can be identified and addressed early on, preventing costly and potentially harmful repercussions. This proactive approach builds trust with users and stakeholders and streamlines the development process. Addressing concerns as part of the development lifecycle reduces the need for major rework or costly fixes later, ultimately saving time and resources.

Furthermore, the trinity fosters a culture of responsible innovation within the organization. By integrating ethical considerations into daily decision-making, development teams are empowered to build AI systems aligning with business objectives and societal values. This approach minimizes risks and ensures that AI is harnessed for positive impact, promoting fairness, transparency, and accountability in an ever-evolving technological landscape.

Technology Obsolescence and Risks

AI is touted as the Fourth Industrial Revolution, from optimizing logistics to personalizing customer experiences. AI systems are becoming integral to modern operations, moving up the value chain, and becoming part of decision support systems. However, AI systems require ongoing maintenance to function optimally like complex machinery. The periodic assessment of the potential obsolescence of the technology used in developing AI solutions is a major factor in system upkeep. Dated system libraries and frameworks can lead to security incidents, cause incompatibilities with newer systems, and create a roadblock to leveraging the latest advancements in AI research. For the AI systems and processes

CHAPTER 21   ALIGNING AI GOVERNANCE, AI DEVELOPMENT LIFECYCLE, AND SYSTEMS DEVELOPMENT LIFECYCLE PROCESSES

described in this book – a mix of those in production and development – it is essential for future-proofing to examine the software and hardware that support AI systems and to determine whether they face the risk of becoming outdated and unsupported. That is the critical, and very likely the most neglected, part of system maintenance.

Imagining an AI solution is like imagining a car. Not providing the required maintenance on a vehicle driven a certain number of miles will cause it to fall apart because the manufacturer's required maintenance schedules were ignored. Keeping an AI solution performing well isn't something you fill a tank of gas with and pay tolls for; it takes consistent, continuous support and scheduled preservation.

If the fast-moving technology world does not allow time for even the most advanced AI systems to stay current, it can compound the issue when those systems are not properly maintained. Perhaps one way to think about this is to consider that a neglected AI system might not just be cutting-edge when it's still new and shiny, but it is even less cutting-edge as time passes and the world continues to move very quickly. In AI systems, it seems the world will maintain the status quo for the foreseeable future. The quick pace of technology makes it so that something revolutionary just a few years ago is probably already a thing of the past today. Systems built on old technologies are more likely to be vulnerable to various threats. Indeed, using an obsolescent technology like the "old facial recognition library in our fictional system" almost ensures that some portion of the training data it was built on is more easily accessible than it would be if we were using, say, one of the modern SDKs to train it with.

The world of technology is in flux, and an excess of new and powerful methods, tools, and platforms overburdens the innovator with a wealth of possibilities. However, these possibilities and the speed of change pose challenges to the long-term viability of AI as a basis for transformative applications and businesses. *Imagine a company that built an artificial intelligence-based premium customer service application where customers communicate with a live chatbot to perform tasks promptly. The chatbot is*

CHAPTER 21 ALIGNING AI GOVERNANCE, AI DEVELOPMENT LIFECYCLE, AND SYSTEMS DEVELOPMENT LIFECYCLE PROCESSES

AI, and AI is the chatbot. The bot's service is very adequate when it comes to getting the job done, which is why it is, by and large, a good news scenario when it comes to the performance of AI.

Now imagine that this same company, for some reason, is still using a software kit that was used to build that AI and has been deprecated by its creators in favor of a much better, upgraded version that works with the platforms and systems that exist today. If the company doesn't change anything, then what ends up happening is that the service can be performed for some customers who are using certain devices or certain platforms but not for others.

The dynamic field of artificial intelligence is continually innovating, and it's projected to create 97 million jobs and add $2.2 trillion to $4.4 trillion to US GDP by 2030. Yet, those who are failing to keep pace with the times and neglecting the continual assessment of their system's technology stack might miss potentially huge opportunities to improve not only the capabilities and performance of their systems but also their algorithms, processing techniques, and hardware needed to create next-generation AI systems. And, of course, the future of any AI – the "national AI strategy" that so many countries tout – lies only in the hands of the "neural networks of one" who inhabits the creator community.

Zero Trust Security is an option worth considering when it involves defense strategies of your AI systems. It is no secret that we are besieged on all sides by bad actors and increased threat vectors in our infrastructure. This is especially concerning, considering we are in "full swing" of artificial intelligence, which has multiplied the number of adversarial attacks and actors for all the right reasons. Our rapidly advancing, fast-paced world is in a race to keep up with the latest innovations. And if we're not careful – if we allow ourselves to be distracted for even a moment – the technologies meant to keep us in the lead can potentially steer us into uncharted territories. *"Our work is secret; the embers of our campfires are often fanned by the same gusts of luck that so many of humankind's innovations have gotten over the centuries."*

CHAPTER 21   ALIGNING AI GOVERNANCE, AI DEVELOPMENT LIFECYCLE, AND SYSTEMS DEVELOPMENT LIFECYCLE PROCESSES

You ensure their long-term success and value by regularly supporting and maintaining your AI systems. Err on the side of a well-defined systems maintenance program, which is imperative in this rapidly evolving field. You must address technology obsolescence risk and ensure efficient, well-designed, well-governed, and trustworthy AI systems.

Defense-in-Depth Strategies Within the SDLC and AI Development Processes

Within the traditional SDLC process, defense-in-depth involves layering security measures at each stage of development. In the planning phase, this might mean conducting thorough threat modeling and risk assessments specific to the AI project. Security features, such as input validation, encryption, and access controls, are built into the architecture during design and development. Testing includes not only functional testing but also security testing to uncover vulnerabilities. Continuous monitoring and logging post-deployment are essential for detecting anomalies and potential threats.

In the AI development lifecycle process, defense-in-depth takes on additional, vastly different dimensions. The training data becomes a target, requiring measures to detect and prevent data poisoning or manipulation. Model strength is a key concern, with defenses against adversarial attacks becoming a critical aspect of testing and validation. The deployment environment also requires careful consideration, with secure model hosting and access controls to prevent unauthorized use or tampering. Additionally, monitoring the AI model's performance in real-world scenarios is essential to detect anomalies indicating a security breach or malicious intent. In critical AI applications, incorporating human oversight is another layer of defense. Humans can review model outputs, identify potential biases, and escalate concerns when AI decisions seem questionable. This "human-in-the-loop" (HITL) approach adds a layer of judgment and accountability that purely automated systems lack, especially in scenarios with significant consequences. Even after your models have been deployed, the process requires constant vigilance. Iterative model validations ensure that performance hasn't degraded over time due to changes in data distributions or other factors. Monitoring tools can detect anomalies, suspicious patterns, or performance drops as early warning systems for potential issues. By proactively identifying these problems, you can address them before they escalate into major incidents.

Rethinking the Development Lifecycle: Building AI in a New Age

The software development life cycle (SDLC) was the framework in which traditional software development and deployments took form for decades. Defined processes, mitigating controls, focusing on functionalities, and delivering a visible, working next system were the instructions that guided developers from their initial idea to production promotion. They had "stage gates" to ensure that the specified phase of the project was completed before allowing it to progress to the next.

Nevertheless, we must change our development practices to accommodate the increasing grievousness of Artificial Intelligence. AI moves rapidly and is a system that continuously learns from the data given to produce a confident outcome over time. Given this mode of operation, an integrated development system involves iteration, speed, and accuracy, and data is paramount to your AI models, i.e., combining SDLC with AIDLC.

As stated in this book, data fuels your AI systems and models.

CHAPTER 21   ALIGNING AI GOVERNANCE, AI DEVELOPMENT LIFECYCLE, AND SYSTEMS DEVELOPMENT LIFECYCLE PROCESSES

The Data Deluge

In the traditional SDLC, requirements trump everything else. Developers meticulously capture every feature and function the software or system needs to possess. However, in the AIDLC, AI models rely on data to perform and meet process automation objectives. The better the data, the better the model. At the start of the AI development lifecycle, teams must take a data-centric approach. They must identify the data sources and set up datasets to gather information and clean and validate what they pull in, ensuring the data they use is fit for purpose. The outcome part of the AIDLC determines whether the AI models that follow it will be successful. The bedrock of AI development is data; this is not true for traditional software development. AI development is an iterative process where different models are built, tested, and trained on good data. The better models are those that perform certain tasks very accurately (regardless of their complexity), or that can be understood at the operational level. Advances in data quality and quantity have led to increasingly better models and a much lower bar to entering this space. Despite these advances, model performance and the complexity of many AI models that have made them increasingly difficult to understand have raised many questions to researchers – especially as certain models are deployed in extremely consequential contexts.

The Iterative Process and Continuous Learning Loop

The conventional software development life cycle (SDLC) is a straight-lined pathway from one phase to the next. It starts with planning and moves through the solution's development, testing, and deployment to production. After deployment, the straight line returns to the maintenance phase. The maintenance phase starts with the same planning of the MLOps and the AI system's architecture. However, in the AIDLC, the metrics that describe the model's performance are reviewed after the training data is used to train the model. If the model doesn't perform when it gets to the model validation stage (i.e., if it's not accurate or precise enough), it indicates that the model isn't ready for deployment to production. Hence, the model development process and strategy must be reassessed. This continuous learning loop is central to the AIDLC, ensuring the model's performance continuously improves as it encounters new information.

Demystifying the AI Black Box

Artificial intelligence models can be hard to understand, especially those that use deep learning. Often called a "black box," it is not clear how these models yield the results they do. This isn't just an academic concern. When a deep learning system is adopted that has a greater impact on humans, as in the case of self-driving cars or the use of AI in medical diagnosis, it is crucial to explain what the model is doing and the rationale. The AIDLC's job is to help us understand what must be implemented for AI to be explainable, responsible, and trustworthy. That cannot happen if the AI is a "black box."

The Symbolic Dance Between DataOps and ModelOps

The fascinating field of artificial intelligence centers around a complex interaction between data and models. Models learn from data, and data fuels AI models in which they learn and operate. Models, in turn, act as engines, transforming data into insights and actions. However, fostering this intricate interplay requires a robust development lifecycle that seamlessly integrates data management (DataOps) and model management (ModelOps). This section explores how DataOps and ModelOps join forces to orchestrate a successful AI development life cycle (AIDLC). The seamless integration of data management (DataOps) and model management (ModelOps) into the AI development life cycle (AIDLC) is the expression of the meaningful part of an even more complex and bigger set of structures. Let's explore these in detail.

CHAPTER 21   ALIGNING AI GOVERNANCE, AI DEVELOPMENT LIFECYCLE, AND SYSTEMS DEVELOPMENT LIFECYCLE PROCESSES

DataOps

DataOps is the foundation of your AI systems' success and value add. DataOps, in the context of AI strategy, is the methodology that promotes collaboration, automation, and continuous improvement in managing the data lifecycle for AI initiatives. It streamlines the processes of data ingestion, preparation, transformation, and delivery, ensuring that AI models have access to high-quality, relevant data promptly. By automating repetitive tasks, DataOps reduces the risk of human error and accelerates the development and deployment of AI applications. Furthermore, it enables the tracking and versioning data and models, facilitating reproducibility and traceability, which are essential for effective governance and compliance.

In terms of AI governance, DataOps plays a crucial role in ensuring ethical and responsible AI practices. It enables organizations to establish robust data governance frameworks that address data privacy, security, bias, and fairness issues. DataOps tools and processes can monitor data quality, detect anomalies, and mitigate risks associated with AI models. Additionally, DataOps fosters collaboration between data scientists, engineers, and business stakeholders, ensuring that AI systems are aligned with organizational goals and comply with relevant regulations and ethical guidelines.

The AIDLC benefits from the DataOps approach in the following ways:

> **Acquisition and Ingestion** – DataOps ensures that the AIDLC efficiently acquires structured and unstructured data from various sources. DataOps ensures the data ingested into the AIDLC is in an explainable, consistent format.
>
> **Preprocessing and Cleansing** – Existing data is rarely pristine. Indeed, the massive amount of data makes it more difficult to collect error-free. As such, the data must be cleansed, preprocessed, and scanned for missing

values and anomalies using data governance processes and tools. An advantage of the DataOps approach is that it reasonably assures accurate and quality data to train and retrain AI models.

Versioning and Lineage Management – DataOps methodically tracks the source of the data and any changes applied throughout its lifecycle. This process allows developers and MLOps personnel to fully understand the data and how it has been transformed to facilitate adequate and repeatable results further.

ModelOps

ModelOps, in the context of AI strategy, is a set of practices and tools designed to streamline the lifecycle management of AI models, from development and deployment to monitoring and maintenance. It addresses the challenges of scaling AI initiatives by automating repetitive tasks, ensuring reproducibility, and enabling continuous model integration and delivery (CI/CD). ModelOps fosters collaboration between data scientists, engineers, and business stakeholders, ensuring that AI models are aligned with business objectives and deliver value to the organization. It also facilitates monitoring model performance, enabling organizations to detect and address issues such as concept drift or model decay, thereby ensuring AI applications' ongoing accuracy and effectiveness.

ModelOps is crucial in establishing and maintaining responsible AI practices in AI governance. It enables organizations to implement robust model governance frameworks that address bias, fairness, transparency, and explainability. ModelOps tools can track model lineage, monitor model behavior, and detect potential risks associated with AI models. Additionally, ModelOps promotes transparency and accountability by providing detailed documentation and audit trails, ensuring that AI systems comply with relevant regulations and ethical guidelines.

CHAPTER 21   ALIGNING AI GOVERNANCE, AI DEVELOPMENT LIFECYCLE, AND SYSTEMS DEVELOPMENT LIFECYCLE PROCESSES

After securing the data within the AIDLC, the next phase includes ModelOps. ModelOps focuses on the model's entire lifecycle, including creation, training, and operationalizing once it is built and deployed. ModelOps contributes to the AIDLC by performing model experimentation and training, model assessment and validation, and finally, model deployment and monitoring. Tools and frameworks for ModelOps enable data scientists to experiment with various algorithms and configurations. They help facilitate efficient training runs and allow for proper management of different model versions. However, ModelOps is not just about training a model. It's also about rigorously testing it using various methodologies to see how well it performs and whether it produces biases.

ModelOps also considers the model's generalization ability. Can it work well when faced with unseen data? Finally, ModelOps helps carry the model to the production environment, delivering it seamlessly and cost-effectively. It also sets up a performance monitoring framework to see how the model performs. Is it still accurate? Is it, or will it degrade over time?

The Synergistic Power of DataOps and ModelOps

When DataOps and ModelOps work together, they create an outstandingly efficient system that maximizes the value an organization can gain from its artificial intelligence and machine learning initiatives. DataOps keeps everything running smoothly on the data side, ensuring the necessary quality, flow, and data governance, rendering the foundation for developing models. ModelOps keeps everything running smoothly on the model side, ensuring that models are not developed in isolation, but rather part of a continuous, collaborative, and controlled development and deployment process. Working together, DataOps and ModelOps establish an infrastructure, along with standardized procedures and tools, that expresses the organization's operational intelligence layer in the right forms and at the right times to be of the greatest value to the organization's stakeholders.

CHAPTER 21   ALIGNING AI GOVERNANCE, AI DEVELOPMENT LIFECYCLE, AND SYSTEMS
             DEVELOPMENT LIFECYCLE PROCESSES

Red-Teaming and Offensive Security Testing During the AI Development Lifecycle Process

In the realm of AI security, red teaming and purple teaming are necessary practices to fortify the defenses of AI systems. Like ethical hacking, red teaming involves simulating real-world attacks using exploits on AI models to uncover vulnerabilities and biases. A dedicated team of "attackers" employs various techniques, from adversarial inputs designed to fool the model to exploiting flaws in the underlying infrastructure. The goal is anticipating and proactively addressing potential threats before malicious actors can exploit them. By actively trying to "break" the AI system, your security teams can identify weaknesses that adversaries might exploit. Purple teaming, on the other hand, bridges the gap between offense and defense. Fostering collaboration between red and blue teams (defenders) creates a continuous feedback loop for improvement. Red team findings are shared with the blue team, who can strengthen their defenses accordingly. This iterative process ensures that AI systems are resilient against current threats and adaptive to evolving tactics and techniques used by adversaries. Offensive security testing of AI models is essential for several reasons. First, it helps identify weaknesses in model architecture, training data, or deployment environments that could be exploited to manipulate or compromise the AI system. Second, it exposes biases in the model's decision-making process, which could lead to discriminatory or unfair outcomes. Third, it enhances the overall security posture of AI systems, making them more resilient against real-world threats and attacks. Red and purple teaming are not one-time activities; they should be incorporated into your AI systems' ongoing development and maintenance. By continuously testing and refining AI models, you can ensure that your AI-powered solutions are effective but also secure and trustworthy. In an increasingly AI-driven world, offensive security practices are indispensable for protecting sensitive data, maintaining user trust, and ensuring AI technology's responsible and ethical use.

Offensive Security Testing Strategies for AI Systems

A common offensive security testing approach uses the "adversarial attack method," where security testers craft subtle perturbations to input data (images, text, etc.) that can drastically alter the AI model's output, causing misclassifications or incorrect decisions. These perturbations are often imperceptible to humans but can cause the model to misclassify images, misinterpret text, or make erroneous predictions. To help expose blind spots in your model's logic and identify manipulating vulnerabilities, the following adversarial testing techniques can be performed.

Warning *Do not perform these techniques unless you have the adequate knowledge in AI model offensive security testing.*

Evasion Attacks → These attacks aim to bypass or evade AI systems, such as spam filters or malware detectors. Security teams attempt to modify malicious inputs to appear benign, tricking the AI into misclassifying them.

Extraction Attacks → Security teams use exploits to extract sensitive information, such as model parameters or training data, from the AI model.

To defend against adversarial attacks, employ techniques like adversarial training (training the model on adversarial examples), defensive distillation (making the model more robust to perturbations), or gradient masking (hiding the model's internal workings from attackers).

Another strategy of offensive security testing is "data poisoning." Data poisoning is a stealthy but serious threat to the integrity and reliability of AI models. Data poisoning involves deliberately introducing malicious data into the training dataset to corrupt the AI model's learning process. By crafting this tainted data, security teams (attackers) will attempt to cause the model to make biased or inaccurate predictions, ultimately compromising its functionality. The following data poisoning techniques can be performed:

Training Data Contamination: Qualified security teams attempt to contaminate the training data to manipulate the model's learning process by injecting malicious data into the training set and subtly altering the model's behavior over time. The expected testing result is biased or inaccurate predictions that benefit the security personnel (attacker). The test is deemed successful when the biased predictions favor predefined and approved security testing goals.

Detecting and mitigating data poisoning attacks require specialized techniques and continuous monitoring of the training data. Defending against data poisoning requires a multi-layered approach like the following:

- **Data Provenance** – Ensure the authenticity and integrity of training data sources.

- **Anomaly Detection** – Implement algorithms to detect unusual patterns or outliers in the data that may indicate poisoning attempts.

- **Robust Training** – Train models on diverse datasets and use robust learning techniques that are less susceptible to manipulation.

- **Regular Audits** – Training data should be regularly audited to identify potential contamination and retrain models if necessary.

CHAPTER 21   ALIGNING AI GOVERNANCE, AI DEVELOPMENT LIFECYCLE, AND SYSTEMS DEVELOPMENT LIFECYCLE PROCESSES

Threat Modeling is another proactive countermeasure and structured approach to identifying and prioritizing potential threats to AI systems. This defense mechanism involves

- Identifying the critical assets of the AI system, i.e., the model, data, and infrastructure

- Enumerating and brainstorming potential threats to the identified assets, considering technical vulnerabilities and malicious actors

- Analyzing the risks and assessing the likelihood and impact of each threat, prioritizing the most significant risks as part of the risk management strategy

- Developing mitigation strategies and implementing countermeasures to address the identified threats, such as improved security controls, data validation procedures, or model robustness techniques

By proactively identifying and addressing potential threats through threat modeling, your organization can strengthen the security and resilience of your AI systems, minimize the risk of exploitation, and ensure the responsible use of intelligent and transformative automation.

In addition to attacking the model, offensive testing extends to the surrounding infrastructure. This includes "infrastructure penetration testing," where qualified security personnel attempt to exploit vulnerabilities in the systems hosting and serving the AI model. In the AI context, penetration testing is like a stress test for your AI models. This method includes attempting to gain unauthorized access, manipulate the model's parameters, or extract sensitive data. Traditional penetration testing is essential for ensuring the AI system is protected from external threats and maintaining the confidentiality, integrity, and availability of AI-related data. Penetration or Pen-Testing involves performing simulated adversaries trying to "break" the AI system in various ways, from non-evasive to brute force attacks. This can include developing adversarial examples and slightly altered inputs that trick the model into making incorrect decisions. Other methods consist of

- Finding vulnerabilities and running scripts and exploits against those vulnerabilities based on real-world scenarios

- Probing the underlying infrastructure and API endpoints for vulnerabilities that could allow unauthorized access or data extraction

The goal is not just to find flaws but to understand how those flaws could be exploited, providing valuable insights to strengthen the AI system's security and resilience against potential threats.

Aligning AI Development Lifecycle and SDLC – Case Study

Aligning AI governance with the Software Development Life Cycle could be considered essential or best practice. However, my experience with this alignment provided positive results and additional controls from a development, validation, and deployment perspective. The significant realized benefit I received was in the areas of deploying the AI system to production environments. Many controls were relevant and useful, i.e., approvals, segregation of duty prevention processes, stage gates, promotion to production controls, and testing within the SDLC process. I also ensured that ethical considerations, risk mitigation, and responsible AI practices were woven into the system development process, from conception to deployment and post-implementation reviews.

Table 21-1 illustrates the use case and mapping (or alignment) used to integrate AI governance and development with a traditional SDLC process.

CHAPTER 21   ALIGNING AI GOVERNANCE, AI DEVELOPMENT LIFECYCLE, AND SYSTEMS DEVELOPMENT LIFECYCLE PROCESSES

Table 21-1. *AI Governance, AIDLC, and SDLC Integration*

Phase	AI Governance Activity	SDLC Action	Outcome
Planning	Defined AI Governance Framework outlining principles, responsibilities, and risk assessment procedures.	Established project goals and identified potential AI applications.	A clear understanding of ethical boundaries and a roadmap for responsible AI development.
	Conducted initial risk assessment for bias, privacy, security, and unintended consequences.	Defined project scope, requirements, and timeline.	Proactive identification and prioritization of potential risks.
	Performed data ethics review to ensure ethical data sourcing and usage.	Collected and documented relevant datasets for AI model training.	Ensured data aligned with privacy regulations and ethical guidelines.

(*continued*)

Table 21-1. (*continued*)

Phase	AI Governance Activity	SDLC Action	Outcome
Design	Incorporated bias mitigation techniques into the model design (e.g., diverse training data, fairness constraints).	Designed AI model architecture and algorithms.	Reduced the potential for discriminatory outcomes and improved model fairness.
	Built explainability features into the model to enhance transparency.	Designed monitoring mechanisms for tracking model performance and detecting issues.	Enabled stakeholders to understand the reasoning behind AI decisions and ensured accountability.
	Implemented security protocols to protect the model and its data.	Integrated security measures into system architecture.	Safeguarded sensitive information and mitigated potential vulnerabilities.

(*continued*)

Table 21-1. (*continued*)

Phase	AI Governance Activity	SDLC Action	Outcome
Development	Conducted continuous risk assessments as the model evolved.	Developed and trained AI model iteratively.	Proactive identification and mitigation of emerging risks.
	Rigorously validated the AI model to ensure accuracy and fairness.	Tested the model against established performance benchmarks.	Ensured the model met performance requirements and adhered to ethical guidelines.
	Documented all development stages thoroughly.	Maintained comprehensive documentation of code, data, and model changes.	Enhanced transparency, facilitated future audits, and enabled troubleshooting.
Testing and Deployment	Conducted comprehensive bias testing before and after deployment.	Deployed the model into the production environment.	Identified and addressed any discriminatory outcomes before and after deployment.
	Performed security testing to identify and patch vulnerabilities.	Monitored model performance and user feedback.	Ensured the model's security and effectiveness in real-world scenarios.

(*continued*)

CHAPTER 21 ALIGNING AI GOVERNANCE, AI DEVELOPMENT LIFECYCLE, AND SYSTEMS DEVELOPMENT LIFECYCLE PROCESSES

Table 21-1. (*continued*)

Phase	AI Governance Activity	SDLC Action	Outcome
Maintenance	Conducted regular audits to assess ongoing compliance with the AI governance framework.	Regularly updated and retrained the model as needed.	Maintained alignment with ethical principles and ensured the model remained accurate and relevant over time.
	Established incident response protocols to address malfunctions or breaches promptly.	Implemented a feedback loop to incorporate user feedback and learnings into future iterations.	Enabled rapid response to potential issues and fostered continuous improvement of the AI system, ensuring its long-term value and alignment with ethical principles.

AI Development Lifecycle to Traditional SDLC Lessons Learned

Despite the successful process of aligning the AI development process with the SDLC process, I discovered that the traditional development approach still needs more work to capture further other AI development lifecycle components, i.e., model feasibility, model validation, metrics and key performance indicators, and continuous monitoring. While these components are available within the constructs of the AI governance and model risk management framework, the SDLC process will need to "play catchup" to include these elements for comprehensive adaptation and interoperability. Unlike regular software, AI models are not static; they learn and evolve. This requires continuous monitoring and retraining throughout the SDLC, especially post-deployment, to ensure the models remain accurate and reliable. The lessons learned here emphasize the importance of flexibility and adaptability within the SDLC to accommodate the iterative nature of AI development. Another key lesson is the necessity for multidisciplinary collaboration. AI projects thrive when domain experts, data scientists, engineers, and ethicists work together seamlessly. Domain experts provide critical insights into the real-world problems AI aims to solve, data scientists bring their expertise in model building and evaluation, engineers handle the technical deployment, and ethicists ensure the AI system aligns with societal values. This collaboration fosters a holistic approach, balancing technical excellence with ethical considerations.

Finally, the experience has highlighted the importance of robust documentation and explainability. AI models are often complex, leading to potential challenges in understanding their decision-making processes. Detailed documentation of model architecture, training data, and validation results is critical for transparency and auditability. Furthermore, incorporating explainability techniques into AI models allows for better human understanding and oversight, promoting accountability and building trust in AI-driven systems. These lessons emphasize the importance of building functional AI models and models that are transparent, accountable, and aligned with human values.

Getting the SDLC Process Ready for the AI Development Lifecycle Process

Further aligning the traditional SDLC process with the AI development lifecycle process involves embracing a more iterative and adaptive approach. Unlike traditional software development, where requirements are often well-defined up-front, AI models are data-driven and require continuous learning and refinement. This means incorporating feedback loops and retraining cycles throughout the development process into the traditional SDLC. Instead of a linear progression, AI projects often necessitate multiple iterations to optimize model performance, address bias, and ensure alignment with real-world scenarios. Collaboration becomes even more critical in an AI-focused development lifecycle. Bringing together data scientists, domain experts, ethicists, and engineers early on fosters a shared understanding of the problem, the data, and potential risks. This cross-functional collaboration ensures that technical considerations are balanced with ethical and societal implications throughout the development lifecycle. Regular checkpoints and open communication channels enable teams to adapt quickly to new insights or challenges that emerge during model development and deployment.

Another key aspect is the integration of the model validation and approval workflows and continuous monitoring into the SDLC. Again, AI models are not static; they can drift or become biased over time as new data is encountered. Continuously monitoring model performance, regular audits for bias, and retraining with updated datasets are crucial to maintaining the model's accuracy, fairness, and overall effectiveness. The SDLC needs to be designed with these considerations, incorporating mechanisms for ongoing evaluation and improvement throughout the AI system's lifecycle. Of course, the million-dollar question is – Is it worth the time and effort to bridge the gap between the two development processes?

Aligning the traditional SDLC with the AI development lifecycle is undeniably an investment of time and resources. Organizations must adapt their established processes, invest in new tools, and potentially retrain staff. However, the payoff can be substantial. By creating a more seamless integration, companies can reduce the risk of costly errors, accelerate time-to-market for AI solutions, and ensure responsible and ethical development of AI systems. This translates into improved efficiency, reduced costs, and enhanced reputation in a landscape where trust in AI is increasingly important. The benefits extend beyond the technical realm. A tightly integrated approach fosters a culture of collaboration and shared understanding between AI developers and traditional application and software developers. This collaboration ensures that AI projects are not just technically sound but also address real-world needs and consider the broader societal impact of AI. Furthermore, it empowers organizations to proactively respond to changing regulations and emerging ethical concerns, solidifying their position as responsible AI innovators.

The decision to invest in closer interoperability between the SDLC and AI development hinges on your organization's specific needs and priorities. If your organization is heavily investing in AI, the benefits of aligning these processes may be worth the cost and effort. It's a strategic decision that can future-proof AI initiatives, ensuring they are both effective and ethically sound in the long run; however, if your organization is starting on your AI systems journey, my recommendation would be to forgo the effort and prioritize your efforts to more pressing processes and activities and hopefully other SDLC-savvy project managers will develop a universal SDLC to AI development lifecycle integration.

Summary and Thoughts

The convergence of AI development with traditional SDLC processes is an ongoing challenge with significant potential rewards. Lessons learned from integrating the two methodologies have highlighted the need for adaptability, collaboration, and a focus on transparency and accountability. While adapting the SDLC to AI can be complex, the benefits are clear: reduced risks, faster time-to-market, and a more ethical and trustworthy AI system.

CHAPTER 21 ALIGNING AI GOVERNANCE, AI DEVELOPMENT LIFECYCLE, AND SYSTEMS DEVELOPMENT LIFECYCLE PROCESSES

For organizations looking to harness the power of AI, a strategic investment in aligning these processes is inevitable. If you decide to invest in the alignment, start by identifying the key areas where the SDLC can be adapted to better accommodate the iterative nature of AI development. Invest in training to foster cross-functional collaboration between AI teams and traditional software developers. Prioritize clear documentation on model validation, approvals, explainability, and ongoing monitoring to ensure that AI systems are effective and aligned with ethical standards and societal expectations. Embracing this integrated approach paves the way for responsible AI innovation that drives business success and positive societal impact.

Quiz

1. What is the primary goal of aligning AI governance with AI and systems development lifecycles?

    A) Reducing model usage

    B) Enhancing consistency and compliance

    C) Limiting stakeholder involvement

    D) Minimizing documentation efforts

2. Which strategy is NOT typically included in defense-in-depth strategies within AI and systems development?

    A) Multiple layers of security

    B) Single-point failure reliance

    C) Continuous monitoring

    D) Regular audits

CHAPTER 21   ALIGNING AI GOVERNANCE, AI DEVELOPMENT LIFECYCLE, AND SYSTEMS DEVELOPMENT LIFECYCLE PROCESSES

3. What is a key challenge when integrating AI development with traditional systems development lifecycle (SDLC)?

   A) Increased operational costs

   B) Technology obsolescence

   C) Enhanced user experience

   D) Reducing transparency

4. What is the benefit of aligning AI governance with the AI development lifecycle (AIDLC)?

   A) Enhancing operational complexity

   B) Ensuring compliance and ethical standards

   C) Minimizing stakeholder feedback

   D) Limiting model performance

5. True or False: Aligning AI governance with development lifecycles ensures consistency and compliance.

6. True or False: Defense-in-depth strategies rely on a single-point failure approach.

7. True or False: Integrating AI development with traditional SDLC can help address technology obsolescence.

8. Aligning AI governance with AI and systems development lifecycles enhances _____ and compliance.

CHAPTER 21   ALIGNING AI GOVERNANCE, AI DEVELOPMENT LIFECYCLE, AND SYSTEMS DEVELOPMENT LIFECYCLE PROCESSES

9. Defense-in-depth strategies within AI and systems development include multiple layers of security and _____ monitoring.

10. Integrating AI development with traditional SDLC helps address challenges such as technology _____.

CHAPTER 22

AI Through the Lens of Non-technical Business Leaders: Embracing AI with Caution

This brief chapter shares perspectives on navigating the AI landscape and maintaining focus on addressing real-world customer challenges. Business executives must emphasize cutting through the hype by prioritizing a deep understanding of customer needs and pain points. They must detail a strategic approach involving actively listening to, collaborating with, and communicating effectively with clients to determine how AI systems can best unlock practical solutions and drive tangible business value. Furthermore, business leaders can articulate their unwavering commitment to championing AI's safe and responsible use and ensure that AI complements human expertise and decision-making while minimizing potential risks. By adopting this approach, your customers can effectively harness the power of AI to benefit both their businesses and the society they do business in.

CHAPTER 22 AI THROUGH THE LENS OF NON-TECHNICAL BUSINESS LEADERS: EMBRACING AI WITH CAUTION

Understanding the Precautious Adoption of AI Systems from a Business Perspective

AI systems can automate repetitive tasks, analyze vast datasets for valuable insights, and personalize customer experiences. While non-technical business leaders cautiously approach AI adoption, they also see the potential benefits it can bring. They understand that AI implementation requires careful planning, infrastructure investment, and workforce upskilling. Responsible AI governance and clear ethical guidelines are paramount to ensure AI is used fairly, transparently, and without bias.

The rise of artificial intelligence has ignited a mix of fascination and trepidation in the boardrooms of businesses worldwide. While not steeped in the technical nuances of algorithms and machine learning, non-technical business leaders know how AI could revolutionize their industries and potentially disrupt established norms. However, the concerns raised by AI's influence on children and non-technical citizens are urgent and must be addressed. AI has weaved its way profoundly and subtly into the lives of children and non-technical citizens. AI-powered educational tools offer children personalized learning experiences, adaptive tutoring, and engaging content. However, concerns arise over data privacy, potential biases in algorithms, and the risk of over-reliance on technology, potentially hindering the development of critical thinking skills. Additionally, the prevalence of AI in toys and entertainment raises questions about how early exposure to AI may shape children's perceptions and understanding of technology. While AI can offer convenience and efficiency, it also raises concerns about surveillance, data exploitation, and the potential for discrimination. The lack of transparency surrounding AI algorithms can lead to a feeling of powerlessness and mistrust, especially when complex AI systems make decisions affecting their lives.

CHAPTER 22 AI THROUGH THE LENS OF NON-TECHNICAL BUSINESS LEADERS: EMBRACING AI WITH CAUTION

Investing in AI infrastructure, talent acquisition, and ongoing maintenance can strain budgets, especially for smaller businesses. Executives should carefully weigh the potential return on investment against the up-front costs and ongoing expenses, often opting for a phased approach or exploring partnerships to mitigate financial risk. As a business owner, I agree that investing in AI can be a daunting financial undertaking, especially for smaller businesses with limited resources. The potential return on investment is enticing, promising increased efficiency, productivity, and new revenue streams. However, the up-front infrastructure costs, talent acquisition, and ongoing maintenance can quickly add up.

It's crucial to approach AI investment with a cautious yet optimistic mindset. A phased approach, starting with pilot projects or smaller-scale implementations, can help test the waters and assess the value AI can bring to your business. Additionally, exploring partnerships with AI providers or consultants can offer expertise and resources without requiring significant up-front investment.

Ultimately, investing in AI should be based on a thorough cost-benefit analysis, weighing the potential gains against the financial risks. It's essential to have a clear understanding of your business goals and how AI can specifically contribute to achieving them. With careful planning and strategic implementation, AI can become an asset for businesses of all sizes, driving growth and innovation for years.

The impact of AI on the workforce is another key consideration for us. While acknowledging the potential for AI to streamline operations and free up employees for more strategic tasks, we also worry about job displacement and the need for reskilling. We recognize the importance of investing in our workforce's adaptability and preparing them for a future where human-AI collaboration becomes the norm. Perhaps most importantly, we are acutely aware of the ethical implications of AI. We sometimes question the potential for bias in algorithms, the risks of data breaches and misuse, and the impact of autonomous systems on

CHAPTER 22 AI THROUGH THE LENS OF NON-TECHNICAL BUSINESS LEADERS: EMBRACING AI WITH CAUTION

society. While we advocate for responsible AI development, prioritizing transparency, fairness, and accountability in AI systems, the challenge lies in balancing innovation with regulatory requirements, ensuring that AI serves as a force for good, not harm, in the business landscape.

Business process managers see AI as a game-changer in streamlining operations, optimizing workflows, and reducing costs. AI-based automation can handle mundane tasks, allowing human employees to focus on more complex and strategic initiatives. However, process managers must carefully evaluate which processes are suitable for automation and ensure that AI systems are seamlessly integrated with existing workflows. Change management and employee training are crucial to address potential resistance and ensure a smooth transition to AI-powered processes.

Sales and marketing teams leverage AI to understand customer behavior better, personalize marketing campaigns, and improve lead generation. AI-powered analytics can identify patterns in customer data, predict purchasing behavior, and enable targeted marketing efforts. However, these teams must consider ethical considerations when using AI for marketing and sales. Striking a balance between personalization and respecting customer privacy is essential to maintain trust and avoid alienating potential customers.

Finance and IT teams are exploring AI applications for fraud detection, risk assessment, and financial forecasting. AI-powered algorithms can analyze vast amounts of economic data in real time, identifying anomalies and potential risks. However, these teams must ensure the accuracy and reliability of AI models, as errors in financial decisions can have significant consequences. Robust data validation and continuous monitoring are crucial to maintaining the integrity of AI-powered financial systems.

The allure of automation, data-driven insights, and enhanced customer experiences is undeniable, yet we must approach AI adoption with a measured dose of caution. Embracing AI without a technical background might seem daunting, but it's achievable with the right approach. How do I address navigating the AI landscape with caution? In Table 22-1, I'll address several inquiries that we often get from clients and business partners regarding the AI "buzz."

Table 22-1. *AI Inquiries and Concerns from Non-technical Business Leaders*

Inquiry	Response
How can I get past the AI buzz without getting lured into the hype?	Focus on Business Goals. Don't be lured by flashy AI buzzwords. Instead, identify the specific problems you would like AI to solve.
How can I build my AI literacy?	You don't need to be a programmer, but understanding the basics of AI is essential, for example: • Read articles and watch videos explaining AI concepts. • Attend webinars or workshops tailored for non-technical audiences. Understand how this foundational knowledge will help you communicate effectively with other business leaders and your technical team. • Seek help from your AI experts to translate complex technical jargon into understandable language so that you can make investment decisions or better align with your business goals. • Consider consulting with reputable AI consultants or technology partners to improve your AI IQ.

(continued)

CHAPTER 22 AI THROUGH THE LENS OF NON-TECHNICAL BUSINESS LEADERS: EMBRACING AI WITH CAUTION

Table 22-1. (continued)

Inquiry	Response
Will AI take over the business and our customer-facing operations?	While the rise of technology may distance us from traditional human interaction, AI-human collaboration offers a unique opportunity to enhance our capabilities and redefine connection in the digital age. Don't forget the AI–human relationship – AI is a powerful tool that cannot replace human judgment and creativity. AI can perform many job functions; however, it could never replace old-fashioned customer service and interactions, like a good handshake and eye-to-eye contact.
• The AI landscape is constantly evolving: • How do you plan on staying informed of AI strategies for your business model? • How do you plan to keep up with the latest trends, research, and regulations? • Is reading industry publications, attending conferences, and engaging with AI communities an option?	Active Engagement with AI Communities is key. • Participating in AI forums and social media groups dedicated to AI advancements allows for real-time knowledge exchange on emerging trends. • Dedicated Time for Learning. Allocating specific weekly time to read industry publications, research papers, and articles from reputable sources ensures a consistent flow of updated information. • Networking with AI Professionals. Building relationships with AI experts, researchers, and professionals through conferences or workshops provides valuable insights and potential collaborations. • Experimentation with New Tools. Regularly testing and exploring new AI tools provides practical experience and a deeper understanding of their capabilities and potential applications.

CHAPTER 22   AI THROUGH THE LENS OF NON-TECHNICAL BUSINESS LEADERS: EMBRACING AI WITH CAUTION

Understanding the Consumer's Take on AI in Their Daily Lives

Consumers also embrace AI daily, utilizing AI-powered virtual assistants, personalized recommendations, and fraud detection systems. While appreciating AI's convenience and efficiency, consumers are increasingly aware of potential privacy concerns and the need for transparency in data usage. They expect businesses to be up-front about how AI algorithms collect, store, and utilize their data. Building trust with consumers through clear communication and responsible data practices is essential for successfully adopting AI in consumer-facing applications.

As AI continues to permeate society, addressing the needs and concerns of children and non-technical citizens is crucial. This involves ensuring that AI tools used in education are transparent and unbiased and prioritizing the development of essential skills. Education and public discourse on AI are vital for equipping non-technical business leaders with the knowledge and critical thinking skills to navigate an increasingly AI-driven world. Most executives do believe AI has the potential to enhance consumer experiences through personalized recommendations and streamlined interactions. However, it's crucial to prioritize transparency and ethical data usage to ensure consumer trust and avoid potential negative impacts.

Summary and Thoughts

Non-technical business leaders across various functions are cautiously optimistic about the potential of AI to drive innovation and efficiency. We recognize the need for responsible AI governance, ethical considerations, and transparent communication with stakeholders. By embracing AI with caution and addressing potential risks, businesses can unlock the full potential of this transformative technology while building trust with

CHAPTER 22 AI THROUGH THE LENS OF NON-TECHNICAL BUSINESS LEADERS: EMBRACING AI WITH CAUTION

consumers and employees. Business executives should strive to foster value-added AI solutions to excel and cultivate a deeper understanding of AI technologies, encompassing machine learning, natural language processing, and computer vision. Staying abreast of the latest advancements is mandatory to provide innovative solutions. Additionally, specializing in AI governance and model risk management is essential. We must help our customers develop ethical frameworks, mitigate risks through AI risk-management strategies, and ensure regulatory compliance.

Furthermore, service providers should strive to offer comprehensive AI strategy consulting. This entails collaborating with clients to understand their unique business needs, assessing their existing infrastructure and data capabilities, and tailoring AI strategies accordingly. Guiding clients through implementation roadmaps, technology selection, talent acquisition, and change management is key. Thought leadership and education, through publications and workshops, solidify our position as trusted AI advisors further.

Quiz

1. What is a primary concern for non-technical business leaders regarding AI adoption?

    A) Enhanced user experience

    B) Ethical implications

    C) Increased transparency

    D) Reducing operational efficiency

CHAPTER 22   AI THROUGH THE LENS OF NON-TECHNICAL BUSINESS LEADERS: EMBRACING AI WITH CAUTION

2. Why is AI literacy important for non-technical business leaders?

    A) To reduce AI adoption

    B) To make informed strategic decisions

    C) To limit stakeholder involvement

    D) To minimize operational costs

3. Which strategy is essential for building AI literacy among executives?

    A) Ignoring ethical considerations

    B) Promoting responsible AI development

    C) Reducing model transparency

    D) Limiting stakeholder feedback

4. What is a potential benefit of AI adoption recognized by non-technical business leaders?

    A) Increased operational complexity

    B) Driving innovation and efficiency

    C) Reducing transparency

    D) Limiting model performance

5. True or False: AI literacy for non-technical business leaders helps them make strategic, forward-thinking decisions.

6. True or False: Ethical implications are not a concern for non-technical business leaders regarding AI adoption.

CHAPTER 22 AI THROUGH THE LENS OF NON-TECHNICAL BUSINESS LEADERS: EMBRACING AI WITH CAUTION

7. True or False: Building AI literacy among executives promotes responsible AI development and adoption.

8. AI literacy for non-technical business leaders involves understanding AI's capabilities and _____.

9. Building AI literacy among executives helps them make _____ strategic decisions regarding AI adoption.

10. Non-technical business leaders recognize the potential benefits of AI in driving innovation and _____.

CHAPTER 23

Navigating the AI Frontier with This Sales Bible: Sales and Marketing Strategies for AI Governance and Risk Management Solutions

I should have called this chapter the Sales Bible.

Artificial intelligence's rapid advancement and integration across industries has ushered in unprecedented innovation and efficiency. From revolutionizing healthcare diagnostics to optimizing financial markets, AI's transformative impact is undeniable. However, this exponential growth also brings forth a myriad of complex challenges and

CHAPTER 23   NAVIGATING THE AI FRONTIER WITH THIS SALES BIBLE: SALES AND MARKETING STRATEGIES FOR AI GOVERNANCE AND RISK MANAGEMENT SOLUTIONS

ethical considerations. The potential for biased algorithms, unintended consequences, and AI misuse necessitates a robust governance and risk management framework.

This chapter further explores into the importance of establishing ethical and responsible AI practices. It aims to empower sales and marketing professionals with the knowledge and strategies to navigate the AI landscape effectively. By understanding the nuances of AI governance and risk management, professionals can confidently position and sell solutions that address technological needs and prioritize fairness, transparency, and accountability in AI deployment. This chapter serves as a guide to harnessing the power of AI for good while mitigating potential risks, ensuring a future where AI serves as a force for positive change in society.

Understanding the AI Governance and Risk Management Landscape from a Sales and Marketing Viewpoint

AI governance encompasses the policies, processes, and mechanisms that ensure artificial intelligence systems' ethical, responsible, and transparent development and use. It establishes a framework to address potential risks associated with AI, such as bias, discrimination, privacy breaches, and unintended consequences. Risk management, a critical governance component, involves identifying, assessing, and mitigating these risks throughout the AI lifecycle, from design and development to deployment and monitoring. Frameworks like the NIST AI Risk Management Framework provide a structured approach to identifying and managing these risks, offering guidance on establishing governance structures, conducting risk assessments, and implementing mitigation strategies.

The market for AI governance and risk management solutions is burgeoning, driven by the increasing adoption of AI across diverse sectors. Industries with pressing AI governance needs include finance, where

CHAPTER 23   NAVIGATING THE AI FRONTIER WITH THIS SALES BIBLE: SALES AND MARKETING STRATEGIES FOR AI GOVERNANCE AND RISK MANAGEMENT SOLUTIONS

algorithmic decision-making can impact lending, investing, and fraud detection; healthcare, where AI aids diagnostics and treatment but raises concerns about patient privacy and data security; and government, which utilizes AI for public services but must ensure fairness, transparency, and accountability. Other sectors like manufacturing, retail, and transportation also witness a surge in AI adoption, creating a broader governance and risk management solutions market. The competitive landscape in the AI governance and risk management market is evolving rapidly, with a diverse range of players offering specialized solutions. Established technology companies like IBM, Microsoft, and Google provide comprehensive AI platforms with built-in governance features that target large enterprises. Niche startups like Credo AI and Fiddler Labs focus on specific aspects of governance, such as bias detection and explainability, catering to organizations seeking tailored solutions. Consulting firms like Deloitte and McKinsey offer advisory services on AI governance frameworks and risk mitigation strategies, leveraging their industry expertise.

CyberOne, LLC of Plano, Texas, is taking a thought leadership role in providing strategic AI Governance and Model Risk Management consulting services.

- ***Centralized and Unified AI Governance Framework*** → CyberOne has developed a multi-governance approach to its AI governance and model risk management framework. The framework focuses on AI strategy and AI governance alignment, multiple governance and risk management process integration, AI policy development and management, and regulatory compliance change management. The framework addresses AI transparency, AI explainability (XAI), AI Facts, privacy impact assessments, fairness, trustworthiness, responsibility and accountability,

AI audit readiness, third-party (AI) management, feasibility, AI DevOps, validation and approval, deployment, and continuous monitoring. CyberOne also provides strategic consulting in developing AI governance oversight models and Centers of Excellence (CoE).

- ***Cybersecurity and Data Governance*** → CyberOne's AI governance and risk management solution includes publications for secure and safe AI systems and model accuracy using comprehensive data governance and data validation solutions, publishes thought leadership, and offers consulting services on AI governance and risk management.

Industry Coverage → CyberOne provides AI governance and model risk management solutions for mid-size financial services, healthcare, energy, technology, and other regulated industries.

Each player in the market differentiates itself through unique selling propositions. Some emphasize the ease of integration with existing AI systems, while others highlight their advanced explainability algorithms or focus on specific industry verticals. The choice of solution often depends on the organization's needs, risk profile, and budget. The market is characterized by dynamic competition, as vendors continually innovate to address emerging challenges and regulatory requirements.

Understanding the evolving landscape of AI governance and risk management is crucial for sales and marketing professionals. By recognizing the specific needs of different industries and customer segments and analyzing the competitive forces, professionals can effectively position their solutions and tailor their messaging to resonate with potential clients. This knowledge enables them to identify and target high-potential customers, articulate the value proposition of their offerings, and differentiate themselves in a crowded market.

CHAPTER 23   NAVIGATING THE AI FRONTIER WITH THIS SALES BIBLE: SALES AND MARKETING STRATEGIES FOR AI GOVERNANCE AND RISK MANAGEMENT SOLUTIONS

Crafting a Compelling Value Proposition

Articulating a compelling value proposition and business case for AI governance and risk management involves quantifying the tangible and intangible costs associated with unmanaged AI. Financial risks include potential fines and penalties for non-compliance with emerging regulations, legal liabilities from biased or discriminatory AI outcomes, and the costs of rectifying AI-related errors or system failures. Reputational risks are equally significant, as negative publicity surrounding AI incidents can erode customer trust, damage brand image, and lead to loss of market share. By investing in AI governance and model risk management, organizations can mitigate these risks, reducing the likelihood and impact of costly incidents while also demonstrating a commitment to responsible AI use, which can promote AI trust and enhance brand reputation and customer loyalty.

Aligning with regulatory compliance is an important aspect of the business case for AI governance and risk management. As governments increasingly scrutinize AI applications worldwide, organizations face a growing web of regulations and ethical guidelines. Failure to comply can result in significant fines, legal challenges, and operational disruptions. AI governance solutions can streamline compliance efforts by providing tools and processes to assess and mitigate risks, ensuring that AI systems adhere to relevant standards and regulations. This reduces the risk of penalties and positions the organization as a responsible and trustworthy user of AI, enhancing its reputation in the eyes of regulators and stakeholders.

Building trust and transparency is another key pillar of AI governance. Customers, employees, and the public increasingly demand greater transparency and accountability in AI systems. Explainable AI (XAI) solutions, which provide insights into how AI models arrive at decisions, can address these concerns by demystifying AI's "black box" nature. Fairness and accountability measures ensure that AI systems are free from bias and discrimination, fostering trust and confidence in their outcomes.

CHAPTER 23 NAVIGATING THE AI FRONTIER WITH THIS SALES BIBLE: SALES AND MARKETING STRATEGIES FOR AI GOVERNANCE AND RISK MANAGEMENT SOLUTIONS

By implementing such solutions, organizations can demonstrate their commitment to ethical AI practices, build stronger relationships with stakeholders, and differentiate themselves in the marketplace.

The competitive advantages of robust AI governance extend beyond risk mitigation and compliance. Organizations can gain a competitive edge by proactively addressing AI risks and ensuring ethical practices. They can leverage AI to drive innovation, improve operational efficiency, and enhance customer experiences. A strong governance framework can also attract top talent, as skilled AI professionals are increasingly drawn to organizations prioritizing responsible and ethical AI development. By demonstrating a commitment to AI governance, organizations can position themselves as leaders in their respective industries, attracting customers and talent who value transparency, fairness, and ethical AI practices.

In conclusion, the business case for AI governance is multifaceted and compelling. It involves quantifying financial and reputational risks associated with unmanaged AI, aligning with regulatory compliance to avoid penalties and build trust, and fostering transparency and accountability to enhance stakeholder confidence. By investing in robust governance, organizations can mitigate risks and unlock competitive advantages, positioning themselves as responsible and ethical leaders in the AI-driven landscape.

Developing Targeted Sales and Marketing Strategies

Selling AI isn't just about selling technology anymore. It's about ensuring the AI system maintains trust and ethics and that it works equally for everyone. That means understanding who you're talking to. Are they the big-picture CEOs worried about reputation? The risk manager who needs to see the numbers? Or the IT folks who want to know how it fits in and is

CHAPTER 23   NAVIGATING THE AI FRONTIER WITH THIS SALES BIBLE: SALES AND MARKETING STRATEGIES FOR AI GOVERNANCE AND RISK MANAGEMENT SOLUTIONS

compatible with their current setup? Once you know your audience, tailor your message. Don't just talk about features; talk about how AI can solve their problems. Does it help banks avoid bad loans? Does it keep patient data safe? That's what they care about. And remember, there's no one-size-fits-all. Use a mix of approaches → webinars for deep dives, articles for thought leadership, and maybe even some ads to get the word out. Partnering with experts also helps. Show potential clients you're serious about making AI work responsibly. At the end of the day, it's about more than just selling a product. It's about building relationships and becoming a trusted advisor in this rapidly changing AI world.

Identifying Key Decision-Makers → Effective sales and marketing strategies for AI governance and risk management solutions begin with identifying the key decision-makers within an organization. These typically include C-suite executives responsible for overall strategic direction and risk management, risk managers who assess and mitigate potential threats, compliance officers who ensure adherence to regulations, and IT leaders who oversee the implementation and maintenance of AI systems. Engaging these stakeholders early on is crucial for establishing credibility and understanding their pain points.

Engaging C-Suite Executives → C-suite executives are concerned with AI's broader impact on the organization's reputation, financial performance, and long-term sustainability. Therefore, messaging should emphasize the strategic value of AI governance in mitigating risks, ensuring ethical AI use, and maintaining a competitive edge in the market. Demonstrating the return on investment (ROI) of AI governance solutions and highlighting their potential to drive innovation and efficiency can resonate with executives focused on growth and profitability.

Targeting Risk Managers → Risk managers are primarily concerned with identifying, assessing, and mitigating potential organizational threats. When targeting this audience, messaging should focus on how AI governance solutions can help identify and address risks associated with AI, such as bias, discrimination, and unintended consequences. Providing concrete examples of how these solutions can strengthen risk management processes and protect the organization's assets can be highly persuasive.

Appealing to Compliance Officers → Compliance officers ensure that the organization adheres to relevant laws, regulations, and industry standards. Tailoring messaging to this audience involves highlighting how AI governance solutions can streamline compliance efforts, automate reporting, and reduce the risk of penalties or legal challenges. Emphasizing the ability of these solutions to monitor and audit AI systems for compliance can resonate with this risk-averse audience.

Reaching IT Leaders → IT leaders are responsible for the technical implementation and maintenance of AI systems. Messaging targeted at this audience should focus on the practical benefits of AI governance solutions, such as improved model performance, enhanced transparency, and easier debugging. Highlighting how these solutions integrate with existing IT infrastructure and streamline AI development processes can appeal to their technical expertise.

Tailoring Messaging → Effective messaging should be tailored to each customer segment's pain points and priorities. For example, messaging might focus on the risks associated with algorithmic bias in lending decisions or the importance of explainable AI in regulatory compliance when targeting financial institutions. For healthcare organizations, messaging might emphasize the need to protect patient privacy and data security while ensuring the accuracy and fairness of AI-powered diagnostics.

Creating Personalized Content → Developing personalized content that speaks directly to the challenges faced by different industries and customer segments can significantly enhance engagement. This can include case studies, white papers, webinars, or blog posts highlighting real-world examples of how AI governance solutions have been successfully implemented in specific sectors. Demonstrating a deep understanding of their unique challenges, you can position your solution as valuable in addressing their needs.

Utilizing Multiple Channels → A multi-channel approach is essential for reaching a wider audience and maximizing engagement. Leveraging a combination of channels, such as webinars, thought leadership content, targeted advertising, and strategic partnerships, can help you reach different decision-makers at various stages of the buying cycle. Webinars can provide in-depth education and demonstrate the value of your solution, while thought leadership content can establish your expertise in the field. Targeted advertising can reach specific audiences with tailored messaging, while strategic partnerships can extend your reach and credibility.

Leveraging Webinars for Education and Engagement → Webinars are a powerful tool for educating potential customers about the importance of AI governance and showcasing your solution's capabilities. By inviting industry experts, sharing case studies, and offering interactive Q&A sessions, you can create engaging and informative experiences that build trust and generate leads.

Building Thought Leadership Through Content → Creating high-quality content, such as white papers, blog posts, and research reports, can position your company as a thought leader in AI governance. This content can address common challenges, offer insights into emerging trends, and provide practical guidance on implementing effective governance strategies. By consistently sharing valuable content, you can attract and nurture leads, establish credibility, and differentiate yourself from competitors.

CHAPTER 23   NAVIGATING THE AI FRONTIER WITH THIS SALES BIBLE: SALES AND MARKETING STRATEGIES FOR AI GOVERNANCE AND RISK MANAGEMENT SOLUTIONS

Sales Techniques for AI Governance and Risk Management

A consultative approach is key to successfully selling AI governance solutions. This involves actively listening to understand your customer's unique challenges and concerns regarding AI. By educating them about the potential risks and pitfalls of unmanaged AI, you position yourself as a trusted advisor. Back up your expertise by showcasing real-world success stories and hard data demonstrating robust AI governance's financial and operational benefits. This builds credibility and helps customers visualize the positive impact on their bottom line. But the sale doesn't end there. Nurturing long-term relationships through ongoing support, training, and staying ahead of regulatory changes solidifies your partnership and ensures continued success for both your company and your client.

Active Listening → Understand the customer's AI initiatives and pain points. Ask probing questions to uncover their specific concerns and challenges. What are their goals for AI adoption? What risks are they most worried about?

Risk Assessment → Conduct a thorough risk assessment of the customer's existing or planned AI systems. Identify potential vulnerabilities, biases, and unintended consequences. This demonstrates your expertise and helps the customer visualize the problems they might face without proper governance.

Education as Empowerment → Don't assume the customer fully grasps the complexities of AI risk. Share insights on emerging regulations, ethical considerations, and industry best practices. Frame your discussion to empower them to make informed decisions.

Tailored Solutions → Avoid a one-size-fits-all approach. Collaborate with the customer to co-create a governance solution that aligns with their unique needs and risk profile. This could involve customizing existing products, integrating with their current infrastructure, or developing bespoke solutions.

CHAPTER 23 NAVIGATING THE AI FRONTIER WITH THIS SALES BIBLE: SALES AND MARKETING STRATEGIES FOR AI GOVERNANCE AND RISK MANAGEMENT SOLUTIONS

Case Studies as Proof → Share real-world examples of how similar organizations have benefited from AI governance. These case studies should be specific to the customer's industry or sector and demonstrate tangible results such as reduced compliance costs, improved decision-making, or enhanced brand reputation.

The Power of Peer Influence → Testimonials from satisfied customers carry significant weight. Share stories of how AI governance has transformed other organizations, highlighting the positive impact on their operations and bottom line.

Data-Driven Decisions → Leverage data and analytics to demonstrate the ROI of implementing AI governance. Quantify the potential costs of non-compliance, AI failures, or reputational damage. Compare this to the projected savings and benefits of a well-governed AI system.

Beyond the Sale → Ongoing Support: Position yourself as a long-term partner, not just a vendor. Offer ongoing support, training, and education to ensure the customer's AI governance program remains effective and adaptable to changing regulations and technologies.

Staying Ahead of the Curve → Keep the customer informed about evolving AI regulations, industry best practices, and emerging threats. This demonstrates your commitment to their success and helps them stay compliant and competitive in the ever-changing AI landscape.

Building Trust Through Transparency → Be transparent about AI's limitations and potential risks. This fosters trust and credibility, demonstrating your commitment to ethical AI practices. Be up-front about the challenges and work collaboratively with the customer to address them.

Fostering a Partnership → View the customer as a collaborator in their AI journey. Celebrate their successes, learn from their challenges, and continually adapt your approach to meet their evolving needs. This partnership approach can lead to long-term, mutually beneficial relationships.

CHAPTER 23 NAVIGATING THE AI FRONTIER WITH THIS SALES BIBLE: SALES AND MARKETING STRATEGIES FOR AI GOVERNANCE AND RISK MANAGEMENT SOLUTIONS

Value Beyond Compliance → Emphasize that AI governance isn't just about ticking boxes. It's a strategic investment that can unlock AI's full potential, driving innovation, efficiency, and competitive advantage. By focusing on the broader value proposition, you can elevate the conversation beyond risk mitigation and inspire customers to embrace AI governance as a critical component of their overall success.

As previously stated, selling AI governance isn't about pushing a product; it's about partnering with clients to unlock the true value of their AI investments. Successful sales professionals and account managers must take a consultative approach, diving deep into the customer's unique challenges and aspirations. They don't just sell software; they provide expert guidance on navigating the complex landscape of AI risks, regulations, and ethical considerations. This *value-driven approach* involves co-creating tailored solutions, demonstrating ROI through concrete examples and data, and fostering long-term relationships built on trust and transparency. By highlighting the tangible benefits of AI governance – from mitigating financial and reputational risks to driving innovation and efficiency – sales professionals empower customers to make informed decisions that protect their organizations and propel them forward in the AI-driven future.

Sales teams and account managers must educate themselves on selling AI governance and risk management solutions. The increasing adoption of AI across industries and the growing concerns around its ethical and responsible use have made this a high-demand area. Being knowledgeable about the intricacies of AI governance not only builds credibility with potential clients but enables sales teams to identify their specific needs, tailor solutions, and articulate the value proposition effectively. The extent of training required depends on the sales team's existing knowledge and the complexity of the solutions offered. A foundational understanding of AI principles, ethics, and risk management is essential. More in-depth training on specific regulations, industry standards, and a competitive

CHAPTER 23   NAVIGATING THE AI FRONTIER WITH THIS SALES BIBLE: SALES AND MARKETING STRATEGIES FOR AI GOVERNANCE AND RISK MANAGEMENT SOLUTIONS

landscape can be beneficial. The goal is to equip sales teams with the knowledge to have meaningful conversations with clients, address their concerns, and position the solution as a strategic asset.

There are numerous resources for sales teams to enhance their knowledge in this area. Online courses like Coursera and Udemy offer introductory and advanced-level AI governance and risk management courses. Industry publications, white papers, and research reports can provide valuable insights into the latest trends and best practices. Attending industry conferences and webinars dedicated to AI governance can be an excellent way to network with experts, learn about emerging technologies, and stay updated with regulatory developments.

Sales teams and account managers navigating the AI landscape can excel by adopting the aforementioned consultative approach, prioritizing their education on AI, and collaborating with the customer to understand business objectives and automation. By actively listening and asking targeted questions, they uncover the client's unique AI implementation needs and concerns. This allows them to tailor solutions that address technical requirements and align with the organization's risk tolerance and ethical values. By educating clients on unmanaged AI's potential pitfalls and regulatory landscape, sales teams establish themselves as trusted advisors. This educational approach empowers clients to make informed decisions, fosters trust, and sets the stage for long-term partnerships.

Demonstrating the return on investment (ROI) is important in validating the value of AI governance solutions. Sales teams can leverage case studies, testimonials, and data-driven insights to paint a clear picture of the tangible benefits that await their clients. They quantify the potential costs of non-compliance, system failures, or reputational damage, juxtaposing them with the projected savings and operational enhancements achievable through robust AI governance. By fostering transparency and showcasing real-world success stories, sales teams build confidence and create a compelling narrative that resonates with decision-makers. Ultimately, this approach establishes AI governance as a risk mitigation tool and a strategic investment that drives innovation, efficiency, and a competitive edge in the AI-powered era.

When to Bring in the Heavy Hitters

- ***Early in the sales process*** → If the customer lacks a basic understanding of AI governance, introducing the expert early can help establish credibility and build trust.

- ***Complex technical discussions*** → If the conversation leads into technical aspects of AI models, algorithms, or risk assessment methodologies, the expert can provide clarity and answer detailed questions.

- ***Regulatory compliance*** → If the customer needs guidance on specific regulations or industry standards, the expert can offer specialized knowledge and ensure compliance.

- ***Building trust*** → If the customer expresses hesitation or skepticism about AI governance, the expert can address their concerns and provide reassurance through their expertise and experience.

Sales teams can effectively qualify AI governance and risk management opportunities by conducting preliminary assessments to gauge the prospect's level of AI maturity and their awareness of associated risks. This can be achieved through targeted questions about their existing AI initiatives, pain points, and current governance practices. By understanding the client's current state and future goals for AI, sales teams can identify potential red flags or areas of high risk that warrant further investigation. This preliminary assessment serves as a filter, allowing sales teams to focus their efforts on prospects who demonstrate a genuine need for AI governance solutions and have the potential to benefit significantly from them. When a clear need and potential for value are established, should the sales team bring in the resident expert to provide in-depth technical and regulatory guidance, thus optimizing their time and resources?

By combining a consultative approach with expert knowledge, sales teams can effectively engage customers new to AI governance, educate them about its importance, and co-create solutions that address their needs and risk profile.

Here is a sample sales pitch and skit for an AI Governance and Risk Management customer call.

Account Manager → *"Hello, John. I understand your organization would like to learn about our AI governance and model risk management advisory services."*

Customer → *Hi, Richard. This is correct. We are implementing an AI Governance and Model Risk Management framework and don't know how to start. Can you give me an overview of your advisory services in this area"?*

(The 30-second elevator pitch) Account Manager – "Absolutely! In today's AI-powered world, our services will help ensure that your AI models and algorithms are well-governed, ethical, compliant, and risk-intelligent. Our tailored AI governance solutions can safeguard your reputation, optimize performance, and drive innovation for your organization."

Customer → *That sounds great. Can you provide me with a bit more details?*

Account Manager → *"I'm more than happy to go deeper, John. Here is a deeper Dive into our Services:*

Comprehensive AI Risk Assessment → *We'll thoroughly audit your current AI systems to identify potential biases, vulnerabilities, and compliance gaps. This will provide a clear picture of your risk profile and pinpoint areas for improvement.*

CHAPTER 23   NAVIGATING THE AI FRONTIER WITH THIS SALES BIBLE: SALES AND MARKETING STRATEGIES FOR AI GOVERNANCE AND RISK MANAGEMENT SOLUTIONS

Customizable Governance Framework Implementation → *Based on the assessment, we'll work with you to design and implement a tailored AI governance framework that aligns with your specific business objectives, risk tolerance, and regulatory requirements. This framework will encompass policies, procedures, and technical controls to ensure responsible and ethical AI use throughout your organization.*

Bias Mitigation and Fairness Assurance → *We'll help you proactively address algorithmic bias and discrimination concerns by implementing robust testing and monitoring procedures. Our solutions ensure that your AI models are fair, transparent, and accountable, promoting equitable outcomes for all stakeholders.*

Explainable AI (XAI) Solutions → *We'll enhance the transparency and interpretability of your AI models, enabling you to understand and explain the reasoning behind their decisions. This builds trust with regulators, customers, and employees while facilitating easier troubleshooting and improvement.*

Ongoing Monitoring and Compliance → *We'll continuously monitor your AI systems for performance, accuracy, and compliance with evolving regulations. Our proactive approach ensures that your organization stays ahead of the curve and avoids costly penalties or reputational damage.*

Training and Education → *We'll provide comprehensive training for your team to ensure they understand the importance of AI governance and are equipped to implement and maintain the framework effectively. This fosters a culture of responsible AI use across your organization.*

Can I go into the value proposition of our services?"

Customer → *Sure, please do. However, please understand that we are a value-driven organization and use metrics to determine feasibility and value to the business."*

Account Manager → *Wow! You are reading my mind. Let's discuss our value proposition in detail.*

- *We mitigate financial and reputational risk by proactively addressing AI risks, protecting your organization from costly fines, legal liabilities, and negative publicity.*

- *By establishing a strong governance foundation, you can confidently leverage AI to drive innovation, optimize processes, and gain a competitive edge. This will drive and balance innovation, efficiency, and business goals.*

- *By demonstrating ethical AI practices and providing transparent explanations, you build and enhance customer trust, loyalty, and brand reputation.*

- *Streamline Compliance? By helping you stay ahead of evolving regulations with our compliance framework and models, you avoid costly penalties and demonstrate your commitment to responsible AI use.*

- *Empower Your Team? By providing comprehensive training, you can equip your employees with the knowledge and tools to effectively manage AI risks and contribute to the success of your AI initiatives.*

Taking this approach, our solutions can align with your values and business requirements. I understand that navigating the world of AI can be complex, but rest assured, we're here to make it simpler and safer for you. Would you be open to scheduling a follow-up meeting where we can dive deeper into how our AI governance solutions can specifically address your organization's needs and priorities? I'd happily connect you with our resident AI expert to answer any technical questions."

Customer → "Yes, that will be great."

Account Manager → "Fair enough. I will send you a calendar invite for an hour of your time."

CHAPTER 23   NAVIGATING THE AI FRONTIER WITH THIS SALES BIBLE: SALES AND MARKETING STRATEGIES FOR AI GOVERNANCE AND RISK MANAGEMENT SOLUTIONS

In the evolving landscape of AI, a targeted and comprehensive selling strategy for AI governance and risk management solutions is essential, as you can see in the skit and conversation with the customer. It goes beyond merely promoting a product; it's about establishing a trusted partnership with clients, guiding them through the complexities of AI ethics, compliance, and risk mitigation. By understanding the unique challenges different industries face and tailoring messaging accordingly, sales teams can demonstrate the tangible value of these solutions. This builds trust and credibility and positions AI governance as a strategic investment that empowers organizations to harness the full potential of AI while minimizing risks and ensuring responsible, ethical use.

Measuring and Optimizing Sales and Marketing Performance

This chapter section demonstrates into how to measure and refine your sales and marketing strategies for AI governance solutions. By tracking the right metrics, gathering feedback, and embracing continuous improvement, you'll ensure your efforts translate into tangible results for your customers and your business. It's a data-driven journey toward maximizing your value to the AI landscape. There are twelve key points that I would like to share that will help you establish KPIs and metrics for building an effective process for optimizing and measuring the success of your sales and marketing program, especially for AI governance and model risk management.

1. Before embarking on any sales and marketing campaign, it's imperative to establish clear, measurable goals. These goals should align with the overall business objectives and be specific, measurable, achievable, relevant, and time-bound

(SMART). Whether it's increasing lead generation, improving conversion rates, or boosting brand awareness, defining clear goals provides a roadmap for success and allows for accurate performance tracking.

2. Once goals are set, identify the key performance indicators (KPIs) that will be used to measure progress. These KPIs should be directly linked to the defined goals and provide actionable insights into the effectiveness of sales and marketing efforts. ***KPIs include lead generation, conversion rates, customer acquisition cost (CAC), average deal size, and customer lifetime value (CLV).***

3. Invest in reliable tracking systems and analytics tools to monitor and measure performance across various channels and campaigns. These tools should provide granular data on website traffic, lead sources, conversion rates, customer engagement, and other relevant metrics. The data collected should be accessible, accurate, and up-to-date to ensure informed decision-making.

4. Establish a regular cadence for reviewing and analyzing performance data. This could be weekly, monthly, or quarterly, depending on the nature of the campaigns and the velocity of the market. Analyze the data to identify trends, patterns, and outliers. Look for correlations between specific marketing activities and sales outcomes. This analysis will provide valuable insights into what's working and what's not.

5. Don't be afraid to experiment with different approaches. Conduct A/B testing on various marketing elements, such as ad copy, landing pages, email subject lines, or call-to-action buttons. This allows you to compare the performance of different versions and identify the most effective strategies. Continuously test and refine your approach to optimize results. (A/B testing is creating two (or more) versions of a marketing asset, such as a webpage, email, or advertisement, with slight variations (e.g., different headlines, images, or call-to-action buttons). Data-driven decisions, continuous improvement, and innovation are key to maximizing ROI and gaining a competitive edge in the market. A/B testing and experimentation can be applied to AI governance sales and marketing by testing messaging, landing pages, email campaigns, sales pitches, and pricing strategies to optimize performance and increase conversions.

6. Gather feedback from customers throughout the sales and marketing process. Conduct surveys, interviews, or focus groups to understand their perception of your brand, products, and services. This feedback can uncover valuable insights into areas for improvement, customer preferences, and unmet needs. Use this information to refine your messaging, target specific segments, and enhance the customer experience.

7. Ensure close collaboration and alignment between sales and marketing teams. Share insights, data, and feedback regularly to ensure a cohesive approach.

Establish shared goals and KPIs to foster a sense of shared responsibility and accountability. By working together, sales and marketing teams can optimize lead generation, nurturing, and conversion processes for maximum impact.

8. The AI landscape constantly evolves, so your sales and marketing strategies must adapt accordingly. Stay informed about emerging technologies, regulatory changes, and market trends. Monitor competitor activities and adjust your approach to maintain a competitive edge. Be agile and willing to experiment with new tactics to stay ahead of the curve.

9. Utilize automation and AI-powered tools to streamline and optimize your sales and marketing processes. This can include email automation, lead scoring, chatbots, and predictive analytics. Automating repetitive tasks and leveraging data-driven insights can free up valuable time for your team to focus on strategic initiatives and build customer relationships.

10. Recognize and celebrate successes, both big and small. This helps to maintain team morale and motivation. At the same time, be willing to learn from failures and setbacks. Analyze what went wrong, identify areas for improvement, and adjust your strategies accordingly. Embrace a culture of continuous learning and improvement to drive ongoing success.

11. Provide ongoing training and development opportunities for your sales and marketing teams. This can include workshops, conferences, online courses, or mentorship programs. By investing in their professional growth, you empower them with the latest knowledge and skills, enabling them to stay ahead of the curve and deliver exceptional results.

12. Encourage a data-driven culture within your sales and marketing teams. Emphasize the importance of using data to inform decision-making, measure performance, and identify areas for improvement. By embracing a data-driven approach, you can continuously optimize your strategies and achieve sustainable growth.

Measuring Sales Performance

Sales teams in the AI governance and risk management space must focus on key performance indicators (KPIs) that reflect the unique nature of their offerings. Traditional sales metrics like revenue generated, number of deals closed, and average deal size remain important, but they should be supplemented with more specialized metrics. These can include the number of successful proof-of-concept (POC) implementations, customer satisfaction scores related to risk mitigation, and reduced AI-related incidents reported by clients. Sales teams should also track the time it takes to move prospects through the sales funnel, identifying bottlenecks and areas for improvement. Analyzing win-loss data and gathering feedback from prospects who did not convert can provide valuable insights for refining sales strategies and messaging.

CHAPTER 23   NAVIGATING THE AI FRONTIER WITH THIS SALES BIBLE: SALES AND MARKE-
TING STRATEGIES FOR AI GOVERNANCE AND RISK MANAGEMENT SOLUTIONS

Optimizing Marketing Performance

Marketing teams play a key role in generating leads, building brand awareness, and educating the market about the importance of AI governance and risk management. To measure marketing effectiveness, teams should track website traffic, lead generation rates, and the conversion of leads into qualified opportunities for the sales team. Content engagement metrics, such as time spent on a page, social media shares, and email open rates, can provide insights into the resonance of marketing messages. Measuring marketing campaigns' return on investment (ROI) is as important as assessing the cost of acquiring leads and customers against the revenue generated. Marketing teams should also monitor brand sentiment and awareness through social listening and surveys, ensuring their efforts effectively position the company as a thought leader in AI governance.

Summary and Thoughts

Artificial intelligence's hasty adoption across industries transforms business landscapes and introduces new risks and ethical concerns. In this chapter, I aim to equip sales and marketing professionals with the knowledge and strategies to successfully navigate the burgeoning market for AI governance and risk management solutions. It begins by comprehensively understanding the AI governance landscape, including key definitions, target markets, and competitive forces.

To effectively sell these solutions, I must emphasize crafting a compelling value proposition that speaks to the financial and reputational risks associated with unmanaged AI. The chapter guides readers through developing targeted sales and marketing strategies, incorporating consultative selling, demonstrating ROI, thought leadership, and content marketing. It accentuates the need to continuously measure and optimize sales and marketing performance to ensure sustained success in AI governance and model risk management.

CHAPTER 23   NAVIGATING THE AI FRONTIER WITH THIS SALES BIBLE: SALES AND MARKETING STRATEGIES FOR AI GOVERNANCE AND RISK MANAGEMENT SOLUTIONS

Quiz

1. What is the key focus of sales and marketing strategies for AI governance and risk management solutions?

    A) Minimizing customer education

    B) Building trust and demonstrating ROI

    C) Reducing transparency

    D) Limiting stakeholder engagement

2. Which strategy is NOT recommended for understanding the market landscape for AI governance solutions?

    A) Conducting market research

    B) Ignoring competitor analysis

    C) Identifying customer pain points

    D) Understanding regulatory trends

3. What is essential for crafting a compelling value proposition for AI governance solutions?

    A) Reducing customer education

    B) Highlighting unique benefits and ROI

    C) Minimizing documentation

    D) Limiting transparency

CHAPTER 23 NAVIGATING THE AI FRONTIER WITH THIS SALES BIBLE: SALES AND MARKETING STRATEGIES FOR AI GOVERNANCE AND RISK MANAGEMENT SOLUTIONS

4. What is a key element in developing targeted sales strategies for AI governance solutions?

    A) Ignoring customer feedback

    B) Tailoring solutions to customer needs

    C) Reducing stakeholder engagement

    D) Minimizing transparency

5. True or False: Building trust and demonstrating ROI are crucial for sales strategies in AI governance solutions.

6. True or False: Crafting a compelling value proposition involves highlighting the unique benefits and ROI of AI governance solutions.

7. Sales and marketing strategies for AI governance solutions focus on building trust and demonstrating _____.

8. Understanding the market landscape involves conducting market research and identifying customer _____.

Glossary

Please note: At the time of this writing, AI governance is a rapidly evolving field. This list is not exhaustive, and new terms and tools emerge frequently.

A

- Accuracy → The percentage of correct predictions made by an AI model.
- Activation Function → Mathematical function in a neural network that determines if a neuron 'fires,' passing information forward.
- AI Ethics → The field examines moral principles and guidelines for the responsible development and use of AI.
- Algorithm → A set of instructions for a computer, particularly in solving a problem or performing a task within AI systems.
- Artificial General Intelligence (AGI) → A hypothetical AI that matches or exceeds human intelligence across any intellectual task.
- Artificial Narrow Intelligence (ANI) → AI that excels in a specific domain (e.g., image recognition, chess playing).

GLOSSARY

B

- Backpropagation → Algorithm used to train neural networks by adjusting weights based on errors.
- Bias → When an AI model's predictions systematically favor certain outcomes, often due to biases in the training data it learned from.
- Big Data → Large and complex datasets often power AI model training.

C

- Chatbot → AI program that simulates a conversation with humans, often for customer service.
- Classification → The AI task categorizes data into predefined classes (e.g., image classification).
- Cloud Computing → Using remote servers over the Internet for processing and storing data, often key to AI deployment.
- Computer Vision → AI field enabling machines to extract meaning from images/videos.
- Convolutional Neural Network (CNN) → A type of neural network particularly powerful for image and video processing.

D

- Data Augmentation → Increasing the amount and diversity of training data by applying transformations (e.g., rotations, flips) to existing data.
- Data Labeling → Adding tags or metadata to data to make it usable for training AI models.

- Data Science → The field of extracting insights from data, often interlinked with AI.
- Deep Learning → A type of machine learning using neural networks with multiple layers.

E
- Explainable AI (XAI) → Methods to help understand how AI models arrive at their decisions, which are crucial for fairness and trust.

G
- Generative AI → AI models capable of creating new content like text, images, music, or code.
- Gradient Descent → An algorithm used to optimize the parameters of an AI model during training to minimize error.

L
- Large Language Models (LLMs) → Powerful AI models trained on massive datasets can generate human-like text and perform various language tasks.

M
- Machine Learning (ML) → A subset of AI where algorithms learn patterns from data without explicit programming.

N
- Natural Language Processing (NLP) → The field of AI focused on enabling computers to understand and process human language.
- Neural Network → An AI model inspired by the brain's structure, consisting of interconnected artificial neurons.

GLOSSARY

O

- Overfitting – Overfitting occurs when an AI model fits too closely to training data and fails to generalize to unseen data.

R

- Recurrent Neural Network (RNN) → Neural network type designed to process sequential data (e.g., text, time series).

- Regression → AI task of predicting a continuous numerical value (e.g., house prices).

- Reinforcement Learning (RL) → AI, where an agent learns through trial and error, receiving rewards or penalties for actions in an environment.

T

- Training Data → The dataset used to teach an AI model.

- Transfer Learning → Using knowledge from a pre-trained AI model on a related task often requires less data.

GLOSSARY

Chapter 1 Quiz Answers

1.
   I. Balancing Governance and the Lack of Government Oversight
   II. Balancing Risks and AI Innovations
   III. Instituting AI Ethics
   IV. Worldwide coordination of AI control standards is a global issue that affects multinational firms.
2. A, B, D, E
3. B, E
4. False
5. B
6. A, B
7. E- Every and all AI actors are responsible for effective AI Governance and Risk Management
8. False – AI Governance isn't focused on reducing headcount.
9. Have a plan for a comprehensive AI Governance Structure and Strategy
10. E – AI governance and model risk management are the go-to sources and strategies to ensure AI ethics, transparency, and risk intelligence.

GLOSSARY

Chapter 2 Quiz Answers

1. True - Your AI strategy shapes the scope and focus of your governance framework. In turn, practical considerations discovered during governance implementation might necessitate changes to your strategy.

2. Training and Education

3. Key Risks customer face when implementing their AI Strategy

    I. Misaligned Goals→ Artificial intelligence projects that have no strategic business objectives or that do nothing to solve real-world problems

    II. Poor Quality of Data Sources → Unreliable, incomplete, or biased data leads to flawed AI models and outcomes.

    III. Lack of Expertise → Inaccurate, insufficient, or prejudiced data results in flawed AI models and results.

    IV. Underestimating Complexity → Oversimplifying the technical challenges, development time, or costs involved

    V. Ethical Oversights → Failing to proactively address issues of fairness, privacy, or accountability.

    VI. Resistance to Change → Limited buy-in from employees or stakeholders due to fear or a lack of understanding.

- VII. Siloed Development → Teams working in isolation without cross-functional collaboration or shared knowledge
- VIII. Failure to pivot or course correct on planned strategies and agile approaches.
- IX. Lack of executive and senior executive buy-in and sponsorship

4. True – Your AI Strategy and Governance Framework Compliments Each Other → Your AI strategy shapes the scope and focus of your governance framework. In turn, practical considerations discovered during governance implementation might necessitate changes to your strategy,

5. Do not harm or injure patients when used in medicine.

6. The four considerations when choosing an AI governance framework are:

   I. Business Goals and AI Strategy Alignment → Your broader business goals should be the foundation of your AI strategy framework.

   II. Flexibility → Select an adaptable framework to keep pace with the ever-changing AI field and your learning journey.

   III. Ethical Considerations → Ensure your AI strategy builds responsible and unbiased AI systems and models.

GLOSSARY

  IV. Comprehensiveness → Ensure your AI strategy includes governance, privacy, and risk management from the ideation stages through the development, validation, deployment, and ongoing monitoring and retraining stages,

7. Evolving Human Interactions

8. False: We must empower the public through **education and transparent communication** to harness AI's full potential while navigating its challenges.

9. Responsible for the AI system's results

10. True - AI governance is about deciding who has the authority to make decisions about AI

Chapter 3 Quiz Answers

1. AI Governance is the framework of principles, policies, **processes, roles**, and technologies that guide the responsible development and use of artificial intelligence (AI) systems.

2. The five AI model risk factors are model development, data quality, cyber/data security, ethics, and model selection.

3. Data Quality, Data Provenance, and **Data Privacy and Security**

4. E - Data Quality Assessment, Dataset Profiling and Bias Detection, Data Cleaning and Preprocessing, and Data Versioning

5. True

6. False - Data science fuels model development & validation

7. For effective feedback loops, data scientists design mechanisms to collect real-world performance data, detect concept drift, and flag cases where the AI might require retraining or human intervention

8. E - Social Responsibility - To protect brand image against ethical disputes

9. AI Model Risk Management (MRM) is a specific subset of AI governance focusing on identifying, assessing, mitigating, and monitoring the risks unique to artificial intelligence models. Comparatively, AI Model Risk Governance (MRG) is the framework of policies, procedures, and organizational structures designed to oversee the responsible use and management of models across enterprise-wide risk functions. It ensures financial, statistical, or AI-based models are developed, validated, used, and monitored to align with an organization's risk appetite and regulatory requirements.

10. True

GLOSSARY

Chapter 4 Quiz Answers

1. C
2. D
3. B
4. A
5. True
6. False
7. True
8. problem
9. explainability
10. stakeholder

Chapter 5 Quiz Answers

1. C
2. C
3. B
4. B
5. True
6. False
7. False
8. traceability
9. risks
10. lineage

Chapter 6 Quiz Answers

1. C
2. C
3. B
4. B
5. False
6. True
7. False
8. leadership
9. data governance
10. value

Chapter 7 Quiz Answers

1. B
2. C
3. B
4. C
5. False
6. True
7. False
8. RACI
9. data
10. audits

GLOSSARY

Chapter 8 Quiz Answers

1. B
2. B
3. A
4. D
5. False
6. True
7. True
8. AI-specific
9. risk assessments
10. risks

Chapter 9 Quiz Answers

1. C
2. D
3. C
4. B
5. True
6. False
7. True
8. stakeholder trust
9. internal and external policies
10. resources and workflows

GLOSSARY

Chapter 10 Quiz Answers

1. C
2. D
3. D
4. A
5. True
6. False
7. True
8. responsible
9. enforcement
10. comprehensive

Chapter 11 Quiz Answers

1. C
2. B
3. C
4. B
5. True
6. False
7. True
8. legal counsel
9. transparency
10. sensitive

457

GLOSSARY

Chapter 12 Quiz Answers

1. B
2. C
3. B
4. A
5. True
6. False
7. True
8. human oversight
9. context
10. ethical

Chapter 13 Quiz Answers

1. B
2. B
3. A
4. B
5. False
6. True
7. False
8. transparency
9. engagement
10. collaborative

Chapter 14 Quiz Answers

1. B
2. B
3. B
4. D
5. False
6. True
7. False
8. renewable
9. longevity
10. efficiency

Chapter 15 Quiz Answers

1. B
2. C
3. B
4. B
5. False
6. True
7. True
8. anticipatory
9. timely
10. post-crisis

GLOSSARY

Chapter 16 Quiz Answers

1. B
2. B
3. B
4. B
5. False
6. True
7. False
8. skills
9. implications
10. informed

Chapter 17 Quiz Answers

1. B
2. C
3. B
4. B
5. False
6. True
7. True
8. innovation
9. license
10. contractual

Chapter 18 Quiz Answers

1. B
2. C
3. A
4. B
5. True
6. False
7. True
8. compliance
9. accountability
10. training

Chapter 19 Quiz Answers

1. B
2. B
3. B
4. B
5. True
6. False
7. True
8. management
9. performance
10. tracked

GLOSSARY

Chapter 20 Quiz Answers

1. B
2. C
3. B
4. B
5. True
6. False
7. True
8. technological
9. scalability
10. integration

Chapter 21 Quiz Answers

1. B
2. B
3. B
4. B
5. True
6. False
7. True
8. consistency
9. continuous
10. obsolescence

Chapter 22 Quiz Answers

1. B
2. B
3. B
4. B
5. True
6. False
7. True
8. limitations
9. informed
10. efficiency

Chapter 23 Quiz Answers

1. B
2. B
3. B
4. B
5. True
6. True
7. ROI
8. pain points

Index

A

Agile methodologies, 42
AI crisis management
    anticipatory risk assessment/
        management, 308, 309
    collaborative incident response,
        311, 312
    communication
        transparency, 311
    continual evolution/vigilance,
        316, 317
    emergency shutdown
        procedures, 310
    ethical considerations, 314,
        315, 318
    forensic analysis/post-crisis
        review, 312, 313
    malfunctioning, 307
    proactive, 307
AI development life cycle
        (AIDLC), 383
    case study, 398
    deployment/
        implementation, 103–105
    DataOps, 391
    flexibility and adaptability, 403
    governance stages, 86
    model approval, 100–103
    model development/testing,
        91–93, 95, 96
    ModelOps, 392, 393
    model review/validation,
        96, 97
    monitoring models, 107, 108
    predefined metrics, 90
    proposals/feasibility, 87–89
    red-teaming/offensive security
        testing, 394–397
    rigorous testing, 85, 86
    risk assessment framework, 109
AI governance
    AI applications, 3–5
    AI lifecycle, 71
    centralized governance
        model, 7, 8
    challenges, 7
    company struggles, 6
    components, 70
    definition, 67–69
    frameworks/tools/
        services, 11–14
    organizations, 15
    strategy experts, 10, 11

INDEX

AI model risk management
  development and deployment process, 151
  governance, 154
  high-level recommendations, 154, 156–158
  KRI, 158, 160–162
  organizations, 151
  preliminary model, 163
  proactive approach, 151
  risk assessment methods, 152, 153
AI Readiness Index (AIRI), 352, 353, 356, 357
AI risk-management strategies, 416
AI systems, impact
  algorithmic efficiency, 298, 299
  challenges, 296
  continuous improvement, 297, 298, 304
  efficiency and sustainability, 303
  knowledge distillation and quantization, 300
  lifecycle, 301
  longevity, 295
  transparency, 302
Artificial intelligence (AI), 1, 307
  building capabilities, 26
  business operations, GPTs, 40, 41
  cost factors/drivers, 60–62
  cultural/social impacts, 50, 51
  foundation/alignment, 23, 24
  frameworks, 54–57
  generative AI, 31–39
  governance/communication/evolution, 27
  GPTs, 30–39
  healthcare provider's, 60
  human labor, 29
  identify AI opportunities, 24, 25
  identifying/managing risks, 44
  implementation, 42, 43, 57, 59
  inclusive design, 48, 49
  intelligent automation, 63
  principled performance, 28
  public awareness/education, 52–54
  responsibility, 46–48
  risks and ethical concerns, 19–23
Audit Committee, 349–350, 361
Auditing AI systems
  addressing exceptions/findings/remediation, 357, 358, 360
  AI model, 343
  AI readiness assessment, 352–357
  audit program/universe, 345–347
  committee, 349, 361
  data, 343
  essential skillsets, 345
  IT audit programs *vs.* AI audit programs, 344

INDEX

organizational goals, 343
process/risk/controls, 348
risk assessment cycles, 350, 351

B

Bias mitigation techniques, 81, 85, 109, 125, 358, 400
Black boxes, 74, 218, 317
"Bleeding-edge" technologies, 56

C

California Consumer Privacy Act (CCPA), 24, 76, 121, 189, 212, 217, 264
Center of Excellence (CoE), 7, 14, 80, 89, 184, 254, 262
Chatbots, 21, 75, 326, 385, 386
Chief Information Security Officer (CISO), 137, 192, 194
Cloud providers, 296
"Compartmentalization" approach, 140
Compliance framework
  ethical principles, 217
  Federal Legislation, 212–217
  history, 210
  hype, 209
  innovation, 224, 226
  metrics, 227
  practical approach, 219–221
  proposed/enacted/emerging AI regulations, 210, 211
  regulatory framework, 218
  surprise gotchas, 222–224
Comprehensive lifecycle assessment (LCA) approach, 301
Computer Vision, 79, 301, 416
Continuous model integration and delivery (CI/CD), 105, 392
"Crawl" approach, 2
Cross-Industry Standard Process for Data Mining (CRISP-DM), 55
Customer acquisition cost (CAC), 437
Customer lifetime value (CLV), 437
Cybersecurity attacks vectors/mitigants, 148–150
Cybersecurity governance, 136–139

D

Data augmentation, 123, 125, 359
Data-driven decision-making (DDDM), 42, 55
Data governance, 26, 115
  bias mitigation techniques, 125
  communication and training, 124
  continuous data governance improvements, 124, 125
  data management/lifecycle governance, 123
  data monitoring, 133
  data privacy/security, 124

INDEX

Data governance (*cont.*)
   foundational principles/
      frameworks, 120, 121
   metadata, 134
   metadata
      management, 125–128
   model governance
      integration, 124
   model quality, 122
   quality management, 134
   responsible AI, 123
   training data, 122
   unified compliance, 128–133
DataOps, 390–393
Data Science, 76–80, 192, 193
Dated system libraries, 384

E

Enterprise architecture (EA)
   AI/business strategies, 375, 376
   business strategy teams, 379
   designing AI landscape, 376
   flow and infrastructure, 374
   implementation and
      management, 377–379
   implementation resources, 373
   integrating AI, 380
   organization's strategic
      goals, 380
Enterprise governance, risk, and
      compliance (EGRC)
   audit, 240, 241
   components, 232
   definition, 232
   GRC framework, 233, 235, 238
      cross-functional
         collaboration, 235
      data protection/governance,
         237, 238
      healthcare, 239
      monitoring/continuous
         improvement, 237
      operational risks, 238
      policies, 234
      policy alignment, 236
      risk assessment/mitigation
         strategies, 236
      risk assessment
         processes, 234
      tooling/workflow, 236
      tools/platforms, 234
      training/awareness, 237
   integrated approach, 232
   integrated GRC approaches, 231
   proactive *vs.* reactive, 233
   risk assessment processes, 239
   risk management, 232
   validate compliance, 242–245
Explainable AI (XAI), 8, 93, 123,
      274, 322, 423
Explainable AI (XAI) Solutions, 434

F

Factsheets, 365, 371
Financial services industries
      (FSIs), 2, 8

Fourth Industrial Revolution, 384
Framework, AI Governance
　building framework, 173, 174
　ethical principles, 180
　reference model, 180
　three-phase process, 175–178

G

General Data Protection
　　Regulation (GDPR), 24, 76,
　　120, 189, 212, 217, 256, 346
Generative Adversarial Networks
　　(GANs), 30
Generative AI, 30–32, 34, 35, 37–39,
　　213, 215
Governance frameworks
　data governance, 120, 121
　establish AI, 114
　　ethics/governance
　　　strategy, 115–119
　　KPIs, 167–169
　　metrics, 164, 166, 167
　　organizational governance
　　　models, 170
　　security governance, 115
Governance, Risk, and Compliance
　　(GRC), 2, 231, 233–242

H

Historically Black colleges and
　　universities (HBCUs), 29,
　　285, 286, 290, 291

Human-in-the-Loop (HITL),
　　94–96, 387
Human oversight
　challenges/risk mitigation, 269,
　　273, 274
　limitations of AI, 267–271

I, J

Identity and Access Management
　　(IAM), 141
Information technology (IT),
　　20, 46, 134–138, 192,
　　220, 232, 325, 344,
　　375, 425
Intellectual property (IP)
　AI-generated creations, 340
　commercialization
　　concerns, 336
　copyrighted materials, 333
　global collaboration, 340
　global standards/norms,
　　338, 339
　human authorship *vs.* machine
　　agency, 334, 335
　legal frameworks/international
　　collaboration, 337
　ownership/AI-generated
　　content, 334
　shared governance, 339
Intelligent automation, 2, 23, 24,
　　28, 59, 63, 325, 379
Intrusion detection/prevention
　　systems (IDS/IPS), 140

IT and cybersecurity governance
  AI models, 138–143
  backup and recovery procedures, 145, 146
  business goals and strategies, 134
  CISO, 137
  implementation, 136
  incident response plan, 143, 144
  risk management practices, 135

K

Key performance indicators (KPIs), 3, 24, 28, 109, 167–169, 402, 440
Key risk indicators (KRIs), 109, 154, 158, 160, 351
Knowledge distillation technique, 299

L

Large Language Models (LLMs), 316

M

Machine learning (ML), 9, 30, 92, 127, 128, 325, 345, 356, 416
Machine Learning Operations (MLOPs), 7, 55, 100, 105, 142, 356, 392

Mean-Time-to-Detection and Remediation (MTTDR), 162
Model inventory
  AI Factsheets, 365
  discovery tools, 369, 370
  factsheets, 366–368, 371
  identification, 366–368
ModelOps, 390, 392, 393
Model risk factors, 73
Model Risk Governance (MRG), 72, 100
Model risk management (MRM), 2, 5
  definition, 72
  drivers, 74–76
Multi-factor authentication (MFA), 139, 141

N

Natural Language Processing (NLP), 9, 79, 132, 295, 301, 324, 325
Neural network, 30, 298, 299, 345, 386
Non-technical business leaders
  AI system adoption, business, 410–412
  consumers, 415

O

Oversight model
  AI governance roles/responsibilities, 192–197, 199–202

470

COE, 184–190
COE management structure, 190, 191
high-level recommendations, 204, 205
structuring board of directives, 202, 203
sustainable AI-driven, 183

P, Q

Personally identifiable information (PII), 139, 260
Policy management
    auditing cycles, 256
    benefits, 251
    business policies, 255, 256
    challenges, 250
    lifecycle, 252–254
    organization's ethical principles, 249
    policy enforcement, 250
Privacy
    AI actors/legal counsel collaboration, 262–264
    data governance, 259
    deployment phase, 261
    design approach, 259
    development phase, 260
    legal counsel, AI actors, 262
    monitoring phase, 261
    technologies/processes, 260
    validation phase, 261

Privacy by Design approach, 259
Privacy-Enhancing Technologies (PETs), 259, 260, 263

R

Regulators, 47, 174, 185, 209, 226, 310, 322, 423
Return on investment (ROI), 62, 87, 88, 327, 425, 431, 441
Role-Based Access Control (RBAC), 141

S, T, U

Sales and marketing strategies
    AI governance and risk management, 420–422
    compelling value proposition, 423, 424
    developing targets, 424–427
    optimizing financial markets, 419
    performance, measuring/optimizing, 436–441
    ROI, 429, 430, 438, 441
    sales techniques
        account manager, 434, 435
        financial and operational benefits, 428, 429
        needs and risk profile, 433
        professionals, 430
        regulatory landscape, 431
        risk mitigation tool, 432

Sales and marketing strategies (*cont.*)
  sales team, 433
  teams and account managers, 431
Secure development lifecycle (SDLC), 142, 153, 253
Shadow AI, 367
Specific, measurable, achievable, relevant, and time-bound (SMART), 436
Stakeholder engagement
  AI myths/strategies, 283, 284
  AI skills gap/diversity crisis, 285
  build public trust, 282
  challenges, 280, 281
  diverse cultural/racial groups, 288–290
  ethical AI development, 278
  HBCUs, 290
  identifying key stakeholders, 278
  mapping, 279
  methods, 279
  Minority-owned AI, 286, 287
  mitigation risks, 278
  public trust, 278
  regulatory compliance, 278
Stakeholders/actors
  AI design, 322, 323
  board of directors/senior leadership team, convincing, 327, 328
  business executives, literacy, 325, 326
  data-driven decisions, 329
  developing skills and knowledge, 321
  end-users, 323, 324, 329
  regulators, 322
System Development Life cycle (SDLC), 383
  AI black box, 390
  AI development process, 387, 388
  case study, 398
  collaboration, 405
  data, 389
  flexibility and adaptability, 403
  iterative process/continuous learning loop, 389
  model validation and approval workflows, 404
  risks, 384–386
  stage gates, 388

V, W, X, Y

Virtual private networks (VPNs), 140

Z

Zero Trust Security, 386

GPSR Compliance

The European Union's (EU) General Product Safety Regulation (GPSR) is a set of rules that requires consumer products to be safe and our obligations to ensure this.

If you have any concerns about our products, you can contact us on

ProductSafety@springernature.com

In case Publisher is established outside the EU, the EU authorized representative is:

Springer Nature Customer Service Center GmbH
Europaplatz 3
69115 Heidelberg, Germany

www.ingramcontent.com/pod-product-compliance
Lightning Source LLC
LaVergne TN
LVHW010332260326
834688LV00036B/679